The Anatomy of Grief

The Anatomy of Grief

How the Brain, Heart, and Body Can Heal after Loss

Dorothy P. Holinger, Ph.D.

Yale UNIVERSITY PRESS
New Haven & London

Published with assistance from the foundation established in memory of Calvin Chapin of the Class of 1788, Yale College.

Yale University Press books may be purchased in quantity for educational, business, or promotional use. For information, please e-mail sales.press@yale.edu (U.S. office) or sales@yaleup.co.uk (U.K. office).

Set in Janson Roman type by Integrated Publishing Solutions, Grand Rapids, Michigan.
Printed in the United States of America.

Library of Congress Control Number: 2020931420
ISBN: 978-0-300-22623-2 (hard cover: alk. paper)
ISBN: 978-0-300-26476-0 (paperback)

A catalogue record for this book is available from the British Library.

10 9 8 7 6 5 4 3 2 1

Always, to Bill

From the brain and the brain alone arise our pleasures, joys, laughter, and jests, as well as our sorrows, pains, and griefs.

—HIPPOCRATES

Great grief is a divine and terrible radiance which transforms the wretched.

—VICTOR HUGO

Contents

Contents

Preface

Books about grief have been written and rewritten, and the genre continues to grow. There are books on counseling, self-help books filled with advice, treatises on theory, religious books offering solace to the bereaved, and memoirs and novels that detail the experience of grieving. Yet despite these increasing numbers of resources, there is a gap: not many books focus on the changes that happen to the human self of the bereaved. *The Anatomy of Grief* is my contribution to help fill that gap. It offers an understanding of what it's truly like after losing a loved one: what the bereaved experience when their brain is grief-stricken, their heart is breaking, and their body is hurting. This book is informed and enriched by my personal life, by my clinical experiences, and by recent scientific research. With science as its foundation and illuminated by the arts, it is a testament of knowledge and hope for anyone who has been affected, at some time in their life, by the pain and emptiness of loss. It is a book for the bereaved.

My focus on the brain began in graduate school as I watched postmortem brain dissections in a neuroanatomy class. Years later, as a faculty investigator studying the postmortem brain, I wondered about the donors. What had life been like for the person whose brain I held? What joys, pleasures, sorrows, laughter, and grief—emotions that Hippocrates, in the fourth century BCE, wrote come from the brain—had they experienced during a lifetime?

In research with living subjects, I studied changes in the electrical activity of the brain using event-related potentials (ERPs), and I measured differences in gray-matter tissue volume using MRI scans in areas of the brain where we hear and understand language.

From my clinical work, it became clear that for some patients in psychotherapy, depression and anxiety covered unexpressed grief. Grief, whether delayed or forbidden, was being expressed through the body as physical symptoms, or through emotions and behaviors that indicated psychological problems. It is not unusual for psychological issues to disguise unaccessed grief, even years after a loved one's death. Forty years after his father's death, a man I knew wept as he remembered how he'd always wanted to tell his father how much he loved him, but never had.

The complexities of research and clinical experiences overlapped, and I found myself standing at the nexus of the physiology and the psychology of grief. This book grew out of that intersection.

This book is unique in that its approach, rather than being prescriptive—offering strategies and ways to overcome grief—is informative and descriptive. I disagree with articles and books stating that grief is predictable and hope to replace this and other myths, misconceptions, and fallacies that may confront and bewilder the bereaved. I stress instead the need for survivors to recognize the tenaciousness of grief by describing what happens to the brain and the body of the bereaved. In doing so, I use examples from clinical and personal narratives; from paleo-archaeology, paleontology, and history; from the performing arts; from research and discoveries in neuroscience, medicine, physiology, and psychology; and from memoirs and other genres of literature. *The Anatomy of Grief*, then, is not a grief memoir, not a self-help book, and not a textbook, but instead offers the reader an understanding of the changeable and unpredictable nature of grief—an understanding that has the power to help ease the heartache of a terrible loss.

Composers, artists, memoirists, novelists, poets, and dramatists have rendered this universal sorrow in music, art, and literature. The genre of grief books includes self-help, religious, and alternative works offering sympathy and advice. Classic and contemporary books describe grief and how to "deal" with it. But

many of them have not focused on the humanness of the bereaved, particularly what the body goes through after the death
of a loved one, and what effect grief has on the human brain.

I wrote this book because of my research on the human brain,
my understanding of grief through my clinical work, my review
of the grief literature, and my personal life. All of these experiences inform and align with my vision of grief.

The book that you are reading contains patient narratives
from my clinical and research experiences as a psychologist. The
names and identifying details of patients have been changed to
protect their privacy. For some patients, I have obtained written
permission to use disguised examples of their grief experience.
For others, I've conflated the details of several persons into one
account.

Throughout the book, terms such as loved one, the beloved,
wife, husband, mother, father, and daughter are frequently used.
Specific relationships, such as mother-daughter and husband-
wife, are discussed, sometimes at length. But no reader is intended to be excluded, whatever the circumstances of their familial relationships may be, and whatever their gender identity
and sexual orientation may be. It can be said generally that in
the United States, a man can be married to his husband, and a
daughter can have two mothers or two fathers. Furthermore,
readers who have adopted children or were themselves adopted
are not meant to be excluded, nor are readers who are single-
parent children and readers who are called significant other and
life partner rather than spouse. There are equivalences among

such variations, and relationships can be strong and emotionally powerful no matter the familial circumstance. Generalities about grief can be made with regard to one's place in the family because where love and death are concerned, family structures share a commonality: there are parents, and children, and siblings. Within family structures there are many, many variations, yet throughout the book, the intention is to include everyone.

In describing the totality of what grief does to the human self, my goal is to give the bereaved the confidence not to fear grief, but to let sadness run its course without familial or societal constraints. For it is then that the nature of grief can change, allowing the intensity of the sorrow to ease and positive memories of the lost loved one to reemerge in the consciousness of the survivor. It is then that a survivor can begin the bitter sweetness of a new life.

Acknowledgments

To the faculty and laboratory colleagues who collaborated with me in my clinical and research endeavors, thank you. And thanks to the directors of those laboratories who provided support, direction, and inspiration. Foremost among them is Donald L. Schomer, who made the Clinical Neurophysiology Laboratory available for my research, and who provided enduring encouragement and support to write grants, papers, and this book. Thanks also to Albert M. Galaburda, who provided new research directions. And many thanks to Martha E. Shenton and the Psychiatry Neuroimaging Laboratory for her collaboration, research acumen, and friendship.

Thanks to my readers: the colleagues and friends in psychiatry, psychology, and medicine at four Harvard hospitals and in private practice, who startled, delighted, and pulled me up short with incisive comments. In particular, tickled thanks to Thomas G.

Gutheil, whose bon mot phrase "cognitive whiplash" pointed out places where critical transitions were needed. Many thanks to David Reisen, whose cogent comments made several chapters clearer. And heartfelt thanks to Michael K. Rees, who once asked me a question I could not answer: "Why don't you write a book?"

My thanks go to those dear friends who rode the bumpy road of book writing with me, including Bobbi Shippey, who made comments that struck and stuck, and Fiona Sinclair, with her smiling, noncommittal nods at whatever oddly nuanced chapter title I happened to mention on the days we met. And thanks to George Santiccioli, who pointed me toward Ariel in *The Tempest*.

There were unexpected deaths while I was writing this book. Colleagues, including laboratory directors and mentors during my doctoral studies at the University of Michigan: William Stebbins, Howard Shevrin, and Naomi Lohr; and Robert McCarley, who was the laboratory director during the tenure of my NIH National Research Service Award fellowship at Harvard Medical School. Others, too, whose lives touched my own in large ways: the wife of a friend, the brother of a friend, and my ex-husband. There was also the pain of encounters with past deaths, my own sorrows, that I hadn't grieved fully until writing this book.

I wish to thank Chris Reid, who, as ideas flowed freely at the very beginning of this project, gave me encouragement. My gratitude as well to Nathan Garland and our phone-conference "huddles," creative times that generated words and ideas that were poignant and illuminating. And my thanks to Barbara J. King, whose laser-focused comments buttressed some of the scholarship and whose suggestions helped with the organizational flow.

I also appreciate the interest and generosity of Jesse Wegman, who provided the extract from his mother's letter to her unborn grandchild, whom she knew she would not live to see, which has become the epigraph for the Epilogue.

Thanks to my agent, Don Fehr. His idea to imagine how the landscape changes from the west coast to the east coast—an image of a plummeting, flat, or precipitous topography—shaped my proposal. Thanks to my editors at Yale University Press: Jean Thomson Black, who championed the idea of the book, and with subtlety steered it in several unexpected but essential directions, and Julie Carlson, who smoothed and polished the book in many places. My thanks also to Jane Hayward, whose experience and talent contributed to this book as she took ideas that floated and placed them in beautiful figures.

For the courage and stalwart commitment of my patients, and for the generosity of all those who participated in my research studies, my heartfelt admiration and thanks go to you.

To everyone who cheered me on from the periphery, like Paul C. Holinger and Derek C. Polonsky, and many others not mentioned, I thank you.

And finally to my husband, Bill, whose patience and talent kept me grounded, who edited the writing, supported me, and loved me: I use my word when feelings are too deep to be uttered—ineffable—to describe my love and thanks.

Introduction
From the Depths

Grief is inextricably bound to love. It looms as large as the loved one we lost, and it is the price we pay for love. As Louis XV, beloved king of France in the eighteenth century, predicted for his country after his death, "Après moi, le déluge." Indeed, after the death of a loved one, the deluge of sadness and heartbreak can flood the grief-stricken. In a bleak world, wet with tears, feelings of loss can seem unbearable.

Sometimes you don't know you're dealing with grief. In the fourth year of my doctoral program, after completing my supervised internships with outpatients, I began a clinical rotation on the psychiatry inpatient unit at the University of Michigan Hospital. An inpatient internship was new for the department of psychiatry, and new to my psychology department, which had agreed to it. It was the first time for all of us.

That first day on the locked unit, it quickly became clear that

I was facing something totally different from outpatient psychotherapy. Mr. G was my first patient. He was assigned to a psychiatry resident for medication and to me for psychotherapy. Rather thin, quite tall, and in his early sixties, he was a farmer from a family of farmers. Mr. G had inexplicably fallen into a deep depression. He said he'd never been depressed before, but one day, totally unexpectedly, his life working with crops became meaningless to him.

After our initial meeting, Mr. G barely spoke, except about his wife and his devotion to her. He became increasingly despondent. Medication wasn't helping. The team, of which I was the newest member, discussed electroconvulsive therapy (ECT). The resident, the attending psychiatrist, and I spoke about the treatment, its potential risks and potential outcomes, with the patient and his wife. They both agreed to the treatment plan.

Without inpatient experience, without having previously worked with any outpatients whose distress registered at this level, and with no experience with ECT, I was concerned. I hoped Mr. G would be all right, that his memory wouldn't be affected. My supervisor suggested I work with Mr. G after each treatment along with the rest of the team, who were also monitoring him.

On the Monday morning after Mr. G had completed half of his prescribed ECT treatments, I knocked and went into his room. He was eating breakfast. He looked up, smiled, and said, "I'm feeling better." He could answer orienting and memory questions with little difficulty. He began to talk about the farm and his plans for the next season. Clearly, he *was* better. The team halted the ECT treatment.

A week later, Mr. G said, "Doc, I'm okay—I woke up with feelings I haven't felt for a long time!" We both laughed at how he had just told me that his libido had come back.

He began to talk of his losses. He told me his younger brother, JJ, had died several months before his depression hit. JJ's unexpected death was the result of an automobile accident; it was a death very different from that of their father, who had suffered a lingering illness.

Mr. G told me that he'd helped his wife care for his father as his father lay dying in the house where Mr. G had been born. Becoming tearful, he described going back to work on the farm the day after his father's funeral. He said that's what his father had done right after his wife, Mr. G's mother, had died—gone right back to work on the farm. Mr. G was eleven at the time. But when JJ died, something different happened. A month after his brother's death, he couldn't face the farm or anything else. He felt awful, numb. He later realized the "awfulness" began around the anniversary of his father's death, which was a month after his brother had been killed.

Several months after his discharge from the hospital, Mr. G and his wife came in for his follow-up appointment. They told us they thought Mr. G had recovered from his depression. The team agreed. Mr. G explained that although the deaths of his father and his brother were different, the pain and loss of them had pulled up his feelings that had been buried around his mother's death. After his mother died, he said his father would not talk about her, and couldn't bear to let him or his brother speak of her, either. But it wasn't only those losses that had made him feel unsteady. It was when he realized that a sudden or lingering death could happen to his wife that he remembered how devastated he had felt after his mother died. Those experiences of grief, along with the thought that his wife could die unexpectedly, just like his brother, had not only paralyzed him, but also pushed him into a deep depression.

Mr. G's grief after his mother died was forbidden. His father would neither talk about her death nor let his sons talk about it. Unaccessed sorrow had built up during his father's illness, an experience that also led him to feel anticipatory grief. After his father died, that anticipatory grief needed to change into an active grieving of his father's loss. But it didn't. He simply followed his father's example. He didn't talk about his father's death and went right back to work on the farm. He remembered rationalizing that death was simply a natural part of life, the model learned from his father.

After his brother's unexpected death, Mr. G said he was flooded with emotions he hadn't felt before. Those unfamiliar feelings intensified around the anniversary of his father's death a month later. It was then that his earlier griefs—forbidden, anticipatory, and delayed—burst through his defense of using work as a distraction and were experienced in full force. He remembered his mother's death, a loss he was forbidden to talk about, and then his father's that he hadn't fully grieved. And he felt overwhelmed with a fear of losing his wife.

Now he understood. The cascade of sorrows needed to be confronted and expressed. Mr. G told me that with his wife's encouragement, he was able to continue to mourn the losses of his parents and his brother, and to talk about his partnership with his wife, as well as their future life together. They were grateful.

In addition to my clinical rotation on the psychiatry inpatient unit, I was part of a research team that studied psychiatric issues, including grief. In the initial interviews for one study, I talked with people whose grief was prolonged and complicated (at that time it was called pathological). I listened to bereaved patients describe how raw and painful their feelings still were, long after the deaths of loved ones. One patient, a woman, had found her four-month-old baby lifeless in the crib, having succumbed to sudden infant death syndrome (SIDS). She clenched her arms tightly across her chest and cried as she told me how this tragedy continued to haunt her. Another patient, a widower, fought back tears as he told me he still couldn't take his wife's clothes out of her closet five years after her death.

Although I hadn't consciously planned it, my clinical and re-search training began to converge around grief. As a clinician, I worked with patients who experienced grief, anxiety, and depres-sion. As a researcher, I studied changes in the electrical activity of the brain using event-related potentials (ERPs), differences in structure and tissue volume using MRI scans in living subjects, and in postmortem brains, I investigated cell-packing density and neuronal size using a microscope.

To study the brain as I did at three different levels of inquiry was like taking apart Russian dolls and discovering one level inside another. The first level was the scalp, where the electrical activity of the brain was measured from electrodes placed on a patient's head. The second level was the brain itself—under the scalp and inside the skull—where brain tissue volumes were measured from MRI scans. And the third level was the postmortem brain—where neurons were measured from slides.

The third level—looking at and measuring human brain cells, neurons that were Nissl-stained a beautiful magenta color—was captivating. I could see differences in the sizes and shapes of neurons, whose shape resembled a pyramid. And I could identify different cell layers in a tissue section because each of the six layers is characterized by a different cell density.

As an investigator involved in National Institutes of Health–funded studies of the brain, I was a long way from my Brown University undergraduate histology class. But it was that class, when we watched a film of moving cells, that had been pivotal for me. The visiting lecturer told us, "Put away your notebooks and watch the movie." This was not animation. The professor was one of the first researchers to record the movement of actual cells. We watched squamous (skin) cells move in sheets toward each other. "But notice," he said, "that as soon as the two sheets meet and bump into each other, they stop." That was what happened with normal cells. Then the film showed something wildly different: squamous cells that moved relentlessly. Sheets of cells ran over and under whatever was in their way. These were aberrant cells that didn't stop. They were cancer cells.

That was the day my interest in human biology—in cells, and in how the body, and then how the mind, work—ignited a passion that would drive my research and clinical career.

Years later, I did more than watch a film of cells moving: I studied them as a researcher. Using a microscope, I viewed and measured cells from the language, hearing, and visual areas of postmortem brains.

Holding a human brain in one's gloved hands, and feeling the heft of its three-pound weight, challenges the perception of self. This organ, with its folds and furrows, gyri and sulci—populated with about a hundred billion neurons—contains all we know, and orchestrates all we do. We play tennis, we listen to music, we read poetry and novels, we plan for the future. With its neurotransmitters, chemicals that send information across synapses from one neuron to another, the brain arbitrates our senses, moods, and behavior. Everything starts with the brain.

In studying the structures and neurons of postmortem human brains, one brain in particular gave me pause. The cause of death for the twenty-five-year-old woman, who had arranged to donate her brain to medical research, was breast cancer. Because of her donation, I was able to measure cells from her visual cortex. (The visual system in humans has cells that fire when, for example, we see a dog, when we recognize that it is a dog, and when we know that it can be petted.) All that this woman had experienced in her short life had been processed by her brain. It was with her brain, as Hippocrates wrote over two thousand years ago, that she felt life's pleasures, pains, griefs, ambitions, and joys. And I held it in my hands.

Long before Hippocrates, even before there were fragmentary drawings of lithe bison and ochre handprints on cave walls, the sorrow of loss existed. Lives lived then were struggles for survival. How to deal with the dead was part of life, as it is still. But what emotions our ancestral hominins may have felt after the death of one of their own we can only imagine. We do have some indication of what archaic life was like from stone tools and fossils that date back over three million years. The use of stone tools in the Paleolithic Period, the "old stone age," is often used to mark the beginning of prehistory. But there were many hominin species, including some australopithecines, that predate this benchmark.

How did our archaic ancestors deal with kin that died? Some of their early ways of dealing with the dead seem to have involved deliberately carrying the body to funeral caches or cairns. And by the time there were burials—the intentional and purposeful burying of the deceased—hominins had developed a sense of death, which coincided with changes in the brain and the body. While the body had long been upright and locomotion was bipedal, the brain had expanded, and the frontal lobe had become larger. It was at this time in evolutionary history, the Middle Paleolithic, that death for hominins became both real and abstract. A symbolic understanding of life and death had developed.

Burials became more complex by the Upper Paleolithic. In the ceremonial burials of the late Upper Paleolithic, the body of the deceased was often painted with ochre and left with decorative grave goods, artifacts, and mementos that were associated with the deceased.

Prehistory came to an end as the ice sheets receded and agriculture began. Civilization and human history were beginning. Burial rituals to remember and honor the dead were magnified. Funeral ceremonies and memorials developed into elaborate rituals, stylized art, and colossal architecture. Yet despite how elaborate funerary practices became, despite how huge the structures to remember the dead grew, and despite how precious was the art contained in tombs, grief was always there. Grief, the universal and timeless sorrow that accompanies death.

Grief is so central to the human experience that it may seem surprising that many non-human animals also grieve. Swans, for example, once paired, have a lifelong bond that, when broken by the death of one, leaves the remaining swan desolate. Elephants are said to grieve longer and more expressively than any other non-human animal. And even reptiles have been observed to behave in ways that look like grief. As Charles Darwin acknowledged, grief is among the emotions that have a universal expression.[1] This deep and powerful emotion cuts across many species.

Research findings in neuroscience, medicine, genetics, physiology, and psychology continue to expand what we know about the human body. As they do, they give us opportunities to study grief and better understand it. Progress in neuroimaging has given us the human brain's connectome, the white matter tracts that map the neuronal network. Knowledge about the human genome continues to increase. We're learning more about the presence of immune cells in the brain—"memory T cells" that protect the brain.[2] And neuroimaging studies have shown how grief affects the brain. These discoveries in science make a difference to those in the throes of grief. To understand how the death of a loved one can affect one's body, heart, and brain makes the grieving process less frightening. Francis Bacon's seventeenth-century aphorism "knowledge is power" rings true. Knowledge won't take away the ache of grief, but it can help the survivor confront the sorrow of loss and answer the agonizing question, "What is happening to me?"

What does happen to a person devastated by loss? There are physical changes, internal and external. Whether death is unexpected or anticipated, it affects the entire self of the survivor. In the brain, the news registers in the amygdala, the guardian of survival, which raises its alarm to "fight, flight, or freeze." We see this when an animal stops what it is doing, becomes hyperalert, then runs away, prepares to fight, or becomes immobilized with fear.

Upon hearing that a loved one has died, disbelief washes over the survivor, whose body shuts down in different ways. The survivor may freeze in shock, seethe in rage against the news, wail in anguish, or become unable to speak. But all attempts to defend against the reality that the loved one no longer lives, and all attempts to shut out the news of death's reality, will eventually fail. When their loved one's life ends, the newly grief-stricken find themselves in a world not just changed, but irrevocably altered.

Grief is a chameleon. It comes in different forms that affect the life and body of the survivor. It can overlap with other forms of grief; it can erupt as anger, return subdued, surge, and recede again. And the time for its tenancy is open-ended. It has its own schedule, and the schedule is different for everyone. To feel sadness, to endure and experience death's anguishing aftermath, and to put the powerful feelings of sorrow into words are what help to make grief change. When it does, it will drift somewhere else, like a cloud, but may be blown back by the strong wind of an anniversary of the death, or by other reminders like a song or a familiar aroma. The acute, crippling sorrow eventually eases as memories—loving, comic, and sad—emerge. When these sweet, bitter, and loving remembrances begin to sustain and enrich the bereaved, their grief becomes ennobled.

PART I

Grief, Described

ONE

Evolutionary Origins

Nothing in biology makes sense except in the light of evolution.
—THEODOSIUS DOBZHANSKY

What happens after a death? In the immediate aftermath, the body of the lost loved one requires preparation and care, while survivors are often plunged into a primal state of intense, debilitating grief. Over time, human cultures have developed rituals to guide the bereaved, and the body of the deceased, through this important transition. The body is prepared according to mortuary customs, which have their origins in ancient times, while a structured progression of funerary rites prepares survivors for the ultimate farewell. Whether secular or religious, these rituals are the last moments that the living will spend with their loved one.

But first, there is heartbreak.

A teenage boy leaves the house in the morning to walk to his bus. Suddenly a van spins out of control, crashing into him. The injured teen dies in the ambulance on the way to the hospital where his mother is working her shift as an intensive care nurse. The emergency room staff at the hospital, recognizing the boy's name, contact his mother. They meet her as she gets off the elevator and try to prepare her for what she is about to confront.

The nurse walks into a small room and sees her son's body covered from the neck down by a white sheet. The door clicks closed behind her. She touches her son's face. Her soft cries change into heaving sobs, breaths coming in gasps. She stands looking at him, stroking his hair, his face. Lowering herself into the chair alongside him, she remembers reading *The Little Prince*, a favorite book for them. And she remembers how it ends. The little prince dies and becomes a star.[1]

For the boy's mother, time no longer has dimension. Yet it continues.

A knock on the door signals an end to the vigil. Pressing her face against her child's, she touches his hair, her sorrowful "tears of lamentation" spilling on the sheeted body.[2] Finally she gets up, whispering, "Goodbye, my dearest son." She opens the door and walks into the hall. The transitory time spent with her son is over. Later, two men will enter the room and lift the shrouded body of the boy onto a stretcher. With the removal of the boy's body, the formal rituals that deal with the newly deceased begin.

When death happens, as it did for this colleague of a friend, the physical acts of handling the remains, and memorializing the person who has been lost, become symbolic rituals of detachment for the bereaved. These customs and rituals usher in human grief. And as the reality of the loss sets in for the survivors, there is the realization that this is the last time they will spend with their loved one.

Part of the Human Condition

In every city, town, and village around the world, in every culture, we can see memorials built by the living to honor the dead. From the smallest of grave markers to the most spectacular of monuments, these memorials document the grief of those left behind.

Some of the longest-lasting structures have been built to honor kings and queens who have died. King Edward I and Queen Eleanor of Castile, the queen consort of England, were a thirteenth-century royal couple described as inseparable until Queen Eleanor's death in 1290, at age forty-nine. Devastated, Edward isolated himself, writing, "Living I loved her dearly and I shall never cease to love her in death." In her memory, he commissioned the design of twelve "Eleanor crosses"—elaborate, tri-level stone monuments with statues of his beloved queen inside and tall crosses on top. Eleanor crosses were erected everywhere that her cortège had stopped for the night as they took her body from where she had died to London. The final stop before the cortège arrived at Westminster Abbey, where Eleanor's body was buried, was at Charing Cross, then the Royal Mews. A Victorian-era replica Eleanor cross, built in 1863, stands today in front of London's Charing Cross train station. Two of the original twelve crosses that Edward built to honor his wife are also still standing.[3]

In India during the seventeenth century, a spectacular monument was built by Shah Jahan, another monarch who shared an enduring love with his queen. Mumtaz Mahal died in childbirth in 1631, leaving the Mughal emperor Shah Jahan bereft. He built the stunningly beautiful Taj Mahal in her honor and her memory. An example of the Mughal empire style of architecture that "combined Indian, Persian and Islamic influences," the Taj Mahal is considered one of the Seven Wonders of the World. Completed in 1653, the intricate ivory-white marble structure still stands as a memorial to a lost love.[4]

For the great majority of us who are not royals, the external symbols of grief will not be as elaborate as those of kings and queens. For a husband who lost his wife after a long illness, a mother whose child died suddenly, or a daughter whose mother didn't survive a medical procedure, the memorial will of course be different. It may be an engraved headstone, or a bronze plaque on a marble columbarium. Whatever is chosen will serve as a public attestation of sorrow at the loss of their loved one.

Do Non-Human Animals Grieve?

Is grief, so powerful in humans, confined to humankind alone? What about other creatures in the animal kingdom? How do members of other species respond to the death of one of their own?

Consider a story told to a colleague by his adult daughter. On a Caribbean vacation, this young woman drove by an iguana standing by another iguana that had died. Over several days driving by the same location, she saw the two iguanas in the same spot—the one still alongside the other as though keeping vigil. Many other instances of animal grief have been reported by researchers. Carl Safina, in his book about what animals think and feel, describes an experience of a friend who owned two bearded dragon lizards. His friend told Safina that when one of her lizards died, "the survivor hardly moved for a couple of weeks, then resumed a more normal level of activity." Safina wonders, "Is it possible that even a lizard might miss a companion?" He questions the idea of "human grief," answering "there is no such thing"—grief does not belong "exclusively" to humans.[5]

Beyond the reports of anecdotal observations in slow-moving reptiles, do other animals show grief? Do birds and non-human mammals grieve after the death of one of their own? Barbara J. King suggests that while there is variation within any species, animals do show grief.[6] From the largest of land animals, the majestic elephant; to the small prairie vole; to the magnificent marine mammals, such as dolphins and whales, animals' behavior changes after the death of one of their own.

A heartbreaking example of grief was observed in a female orca whale after her newborn calf died. Researchers watched for seven days as the mother orca balanced the dead calf on her nose, trying to keep it afloat. The persistence of the mother whale's efforts showed her anguish.[7] Other animals, too, like elephants and chimpanzees, when faced with the death of a group or family member, can show what King describes as "mourning behavior." It has become clear that many animals indicate their grief by changing their eating and sleeping patterns, making distressed vocal sounds, and exhibiting facial expressions that signal their despair.[8]

Birds, too, change their calls when one has died, and the amazing and intelligent corvids—ravens, crows, and magpies, which some call "feathered apes"—show differences in the arrangements of their feathers, which is a known way of communicating fear, arousal, or other emotions.[9] John M. Marzluff and Tony Angell describe how crows aid their sick with the vigilance of their group members, and how they deal with their dead. A crow "funeral" is distinctive: often hundreds, perhaps thousands, of crows will caw and converge at a place where there is a dead crow. The noise can last up to fifteen minutes and is followed by an abrupt silence. Then all fly away in a silent departure.[10]

It is perhaps not surprising that those animals which mate for life show grieving behaviors after the death of a mate. The male prairie vole, minute and monogamous, does not look for another mate if his female vole dies.[11] And swans, who once paired have a lifelong bond, seem desolate when this bond is broken by death.

Of those massive and magnificent creatures, elephants, King writes that they "are a touchstone species for understanding how wild animals grieve."[12] After the death of an elephant mother, a nursing elephant can die of hunger, while an older juvenile can die of heartbreak. Is it not grief when an animal's behavior demonstrates distress upon the death of a mate, an offspring, or a contributing member of the social group? Darwin seemed to think so, for he stated that grief is among the emotions that are expressed universally.[13]

For humans, after a death—in the brief time before the rush of funerary activities—a survivor or survivors may stay beside the body of their deceased loved one. Is there a neurological signal that cuts across species, a prompt, to remain beside the body of one of their own for a time? Perhaps for the living, the act of keeping vigil over the body of the deceased is a way to confront the reality of the death. We are also not alone in the animal kingdom in showing later grieving behaviors—lethargy, loss of appetite, and disturbed sleep.

Is it possible that over millennia, ancient responses to death evolved, eventually encompassing creations of symbolic monuments as beautiful as the Taj Mahal? Before humans emerged, what happened to the body of the deceased? How did *expressions* of grief change over time? Are brains, no matter how different in size and structure, wired similarly when it comes to responding to death? Is grief common to human and non-human primates and other animals?

To begin to answer these questions, let's turn the evolutionary clock back to before our primate ancestors diverged from early mammals. And as startling as the question seems, it's worth asking: was there grief even before there were mammals on Earth?

Table 1. Geologic Time Scale

Era	Period	Epoch	Time (Millions of Years Ago)	Biological Systems, Age, Events
Mesozoic	Triassic	—	251–199.6	Early reptiles
	Jurassic	—	199.6–145.5	Age of Reptiles
	Cretaceous	—	145.5–65.5	— Cataclysmic upheaval
Cenozoic	Tertiary	Paleocene	65.5–54.8	Early mammals
		Eocene	54.8–33.9	Age of mammals
		Oligocene	33.9–23	Early primates
		Miocene	23.3–5.3	Hominids
		Pliocene	5.3–2.6	Hominins, first human ancestors
	Quaternary	Pleistocene	2.6–.012	Archaic humans
		Holocene	.012–present	Beginning of civilization

The information in this table was culled from various open sources, including sources cited in this chapter.

Climate and Plant Life	Representative Animal Life	Technology (Tool Development)
Dry, arid	Small dinosaurs appear	—
First flowering plants	Dinosaurs dominate (small mammals)	—
Flowers/plants — Lengthy time without sunlight	Dinosaurs become extinct; 75 percent of species disappear	—
Photosynthesis returns	Proto-primates	—
Tropical	Prosimians	—
Cooler, dryer	Old/New World monkeys	—
Temperate: savannahs, grasslands	Great Ape lineage	Pre-Paleolithic
Ice Age begins	*Australopithecus, Kenyanthropus*	Lower Paleolithic (stone tools)
Glaciers	Speciation of *Homo* genus (e.g., *H. habilis, H. erectus, H. heidelbergensis, H. neanderthalensis, H. sapiens*)	Middle Paleolithic (flaked tools) — Upper Paleolithic (blade tools)
Ice sheets recede — Agriculture begins: plants, edible crops	*Homo sapiens*	Mesolithic (Bronze Age) — Neolithic (Iron Age)

The Age of Reptiles

The dinosaurs thundered over the Earth's surface for millions of years, arising during the Triassic Period about 240 million years ago (see Table I). Dinosaurs dominated the planet throughout the Age of Reptiles and existed until their extinction at the end of the Cretaceous Period, around 66 million years ago.[14] Herbivorous dinosaurs like the *Diplodocus* and *Brontosaurus*, and millennia later the carnivorous *Tyrannosaurus rex*, lived on diets of plants and prey. It is difficult to imagine, but could a dinosaur, say a *Brontosaurus* who lived in that remarkable world during the Jurassic Period, have shown some change in its behavior when one of its own died?

In the academic world of paleontology, the prevailing view of dinosaurs before 1969 was that these immense animals had "plodded their way to extinction at the close of the Mesozoic world."[15] But when John Ostrom described his discovery of a dinosaur he named *Deinonychus* ("terrible claw") to his colleagues, it caused a paleontological revolution. Ostrom's details of *Deinonychus* as a swift-moving predator, agile and intelligent, crushed the view that dinosaurs were slow and lumbering. Ostrom's discovery led to what Robert Bakker later called a "dinosaur renaissance."[16]

Could the behavior that we see in contemporary reptiles—to return to the iguana and the bearded dragon lizard described earlier—be a way for us to imagine that a dinosaur mother standing by her deceased young dinosaur might experience grief?

Another example of current reptile behavior that seems grief-like has been reported by anthropologist King, and involves a sea turtle's loss of his mate. Honey Girl lived at Turtle Beach, a protected site for endangered turtles on the Hawaiian island of Oahu. When Honey Girl was found dead on the beach, killed by humans, residents erected a memorial by the sea that included a picture of her. What happened next was astonishing. Tourists visiting the memorial encountered another visitor: "A large male turtle hauled himself out of the water and made his way up the beach straight toward the photograph. There he parked himself, in the sand, head oriented toward the image of Honey Girl. Judging a turtle's gaze as best humans can, observers concluded that he stared hard at the picture for hours. Was the male grieving for his mate?"[17]

King considers whether death can result in a complex range of emotions in reptiles, even in an amphibian that is "many evolutionary eons away from us primates, and indeed from any mammal."[18] This turtle certainly appeared to be grief-stricken by the loss of his Honey Girl. King notes, however, that we will never know with certainty if the male turtle mourned. Similarly, we will never know whether the reported changes in behavior that the iguana and the dragon lizard showed were grief-related. The best we can do is to report behaviors that are consistent with an expression of grief. King writes: "Grief can be said to occur when a survivor animal acts in ways that are visibly distressed or altered from the usual routine, in the aftermath of the death of a companion animal who had mattered emotionally to him or her."[19]

And the dinosaurs? Of course we don't *know* if a dinosaur changed its behavior when one of its own died. We have absolutely no idea of dinosaur emotional life, and there is no way to observe dinosaur behavior from fossils. But it is intriguing to think that during the Age of Reptiles, when dinosaurs dominated the planet, their behaviors were similar to those sometimes observed in their modern-day counterparts—to imagine that those extinct ancestors of contemporary reptiles may have responded to the loss of a companion, or to the death of a young dinosaur, with changes in behavior that expressed their sense of loss.

Although it involves ducks and not reptiles, a personal observation of animal grief by Carl Safina seems fitting here. When one duck, of a pair raised from ducklings, died, the other wandered for days, searching and calling for its companion. In considering whether this behavior is an expression of grief, Safina asserts that "grief doesn't require understanding death."[20]

We know human grief. And we observe what appear to be grief-like behaviors in many non-human animals. Is it unreasonable then to suppose that grief itself was present in creatures long before humans walked the Earth?

Evolution of Mammals and Early Primates

It was close to sixty-six million years ago when a massive asteroid, six to seven miles wide, smashed into the planet, near Chicxulub, where the Yucatán Peninsula is located today. The catastrophic impact caused the extinction of the non-avian dinosaurs and wiped out 75 percent of other animal and plant species. The immense crater that resulted measured 110 miles in diameter, and about eighteen miles deep.[21] The object that caused this apocalypse "lofted twenty-five trillion metric tons of debris into the atmosphere," according to scientists at the Los Alamos National Laboratory who have modeled the cataclysmic impact of the massive meteorite, and the effects of the "blowout."[22] This planetary upheaval brought an end to the Cretaceous Period, and ended the Age of Reptiles. The Age of Mammals was about to begin.

The rubble and debris from this calamitous event blocked the sun, and a dark winter followed. When the dust and debris finally cleared, about sixty-five million years ago, allowing the sunlight again to reach the surface, photosynthesis returned. The survivors of the cataclysm, small animals and sturdy plants, began to emerge, thrive again, and evolve. The new era was called the Cenozoic Era and the Tertiary and Quarternary Periods. The Cenozoic Era, which continues to the present day, included seven geochronological epochs: the Paleocene, the Eocene, the Oligocene, the Miocene, the Pliocene, the Pleistocene, and the Holocene. And it was during these epochs that evolution from a common ancestor, apes, flourished into different lineages, branching into the archaic human species, and ultimately into the species of our *Homo* genus.

The Plesiadapiformes were among the earliest primate-like mammals to evolve during the first Paleocene epoch. These small proto-primates were "transitional" mammals, similar in size to squirrels. The few fossils that have been found suggest that they were arboreal. What sounds, what behaviors could these proto-primates have used after one of their own died? Modern squirrels not only squeal when distressed, but have also been reported to "emit pitiful cries" and show vigil-like behavior in standing beside the body of a companion squirrel that has died.[23]

The Plesiadapiformes diversified into early primates, the pro-simians, during the Eocene, the next epoch. The climate in the early Eocene was very warm. The high level of carbon dioxide resulted in a widespread increase in forests and plant life, conditions that in turn made the landscape habitable for a diverse array of mammals. This second epoch was when proto-horses, rhinoceroses, and the first aquatic mammals appeared, and birds, including eagles and pelicans, flew over the planet. As the Eocene ended, temperatures dropped and rainfall increased. The climate became cooler and dryer in the Oligocene epoch that followed.[24] Primates continued to evolve, but it wasn't until the next epoch, the Miocene, that speciation flourished.

The Miocene is called the "golden age of apes" because it was when apes branched into many species. This epoch was warm with a landscape favorable for mammals: grasses grew and expanded into North American prairies, Eurasian steppes, and African savannahs. During the early Miocene, about fifteen million years ago, apes branched off from monkeys and some split into the family of hominids, the Great Apes.[25] It was out of the lineage of the Great Apes that our earliest human ancestors evolved. They were the hominins.[26]

Fast forward several million years on the geochronological

clock, to about four million years ago. It is the sixth epoch, the Pliocene, and members of the *Australopithecus* genus have evolved; they are bipedal, walk upright most of the time, and have smaller canine teeth. The body of *Australopithecus* was different, and its brain would soon be, too. About 3.4 million years ago there was an incomplete duplication in a brain-development gene—SRGAP2— that triggered a "genetic burst" in the *Australopithecus* brain. What happened is that the extra copy of the SRGAP2 gene duplicated, but not completely, allowing neuronal growth to slow down long enough for more brain cells to develop and neuronal density to increase. And with more brain cells, there were more axons—the single fibers that protrude from neurons and carry messages rapidly from one cell to another. Strategic areas of the *Australopithecus* brain—visual, auditory, motor, and premotor regions—expanded.[27]

Raymond Dart discovered the fossil skull of a three-year-old, dated to be around 3.3 million years old, in South Africa in 1924. He named it the "Taung child" and classified the fossil as *Australopithecus africanus,* the "Southern ape of Africa."[28] The Taung fossil is considered to be among the earliest hominins to walk upright because of where its foramen magnum is located. The foramen magnum is a large hole in the skull through which nerves and blood vessels pass from the brainstem to the spinal cord. Where the foramen is located indicates the angle of the head in relation to the spine. In this *Australopithecus* skull, the foramen was toward the front, meaning that the body was vertical: the Taung child walked on two legs.

In the year 2000 in Dikika, Ethiopia, another *Australopithecus* fossil was discovered: the skeleton of a toddler known as "Dikika baby." The fossil, also thought to be around 3.3 million years old, has a complete humanlike column of cervical and thoracic vertebrae: seven cervical and only twelve thoracic vertebrae com-· pared to the thirteen vertebrae found in African apes. What makes this toddler skeleton remarkable is its second-to-last vertebra, T11, the anatomical segment showing the transition to an adaptation for walking upright. The vertebra in the Dikika fossil— along with those of four other known early hominins—show that upright bipedal locomotion was present in hominins at least 3.3 million years ago.[29]

Paleolithic Tools and the Hominin's Sense of Death

The Paleolithic Period, or "Old Stone Age," was the time in prehistory when our human ancestors first made and used stone tools. Previously, the stone tools unearthed from Olduvai Gorge in Tanzania were considered to be the earliest Paleolithic tools; they were used 2.6 million years ago by *H. habilis*, the first of our *Homo* genus.[30] A more recent discovery of stone implements at Lomewki, Kenya (LOM3), however, predates the Oldowan record by almost a million years, dating back more than 3.3 million years. Sonia Harmand and her colleagues suggest that LOM3 tools may have been used by members of *Australopithecus afarensis* and perhaps another species, *Kenyanthropus platyops*, 700,000 years before the first stone tools used by *H. habilis*.[31]

The LOM3 tools were larger and more primitive than those

from the Olduvai Gorge, yet like the Oldowan ones, they were knapped. Knapping is a way of fashioning stones into usable implements that requires both planning and hand-eye coordination. With the nascent tool in one hand, and the shaping stone in the other, contemporary reenactments show the knapper begin to shape the stone, while hearing the striking sounds and feeling the vibrations. The finished stone tool that hominins crafted probably resembled an imagined or remembered image of the tool that was stored in his or her premotor cortex.[32]

It is interesting to wonder whether it is more than coincidence that members of the *Australopithecus* genus could have been making stone tools around 3.3 million years ago, which is approximately when the SRGAP2 duplication in *Australopithecus* occurred. Although no causal link between the two has yet been found, the duplication in SRGAP2 is considered a pivotal genetic event in evolution: an increased number of neurons in the brain meant more neuronal connections, neural circuits, and cortical mass.[33] This genetic event certainly doesn't tell the whole story of what happened during the evolution of the human lineage from non-human primates. But it does give us a sense of how the brain continued to change in *Australopithecus*. As Megan Dennis and her colleagues point out, "the general timing of the potentially functional copies, SRGAP2B and SRGAP2C, corresponds to the emergence of the genus *Homo* from *Australopithecus (2-3 mya)*."[34]

And grief? What effects could genetic changes in the brains of *Australopithecus* have had on their behavior after one of their own died? How would a member of *Australopithecus* have responded to the death of a companion or a young family member?

How the Dikika toddler died is a mystery. But evidence from an examination of the Taung fossil in 1995 led Lee R. Berger and Ron J. Clarke to hypothesize that the Taung child was a victim of a large bird of prey. A decade later, taphonomic evidence (accumulations of faunal and other fossilized decay) and a reexamination of the fossil supported the "bird of prey" hypothesis. The talon damage in the eye sockets of the Taung fossil looked "nearly identical" to the characteristic bone damage found in primates we know have been killed by large, eagle-like birds of prey.[35]

And the mother of the Taung child? Can we imagine her grief when her child was killed by a large bird of prey? We can't know for sure, of course, but observations of behaviors in contemporary primate species could serve as a model. In the wild and in captivity, mothers in several primate species have been reported to tend to their infants' corpses in ways that demonstrate the maternal bond. There are some descriptions of primate mothers that carry their dead infants around with them for months after the death, even after the bodies have mummified. Indeed, the primate mother-child bond has been closely studied, and what reports there are point to a phylogenetic continuity that cuts across primate taxa.[36]

James Anderson and his colleagues have used a comparative evolutionary perspective to study death and dying in primates.

They documented pre-death and post-death behaviors by chimpanzees as they responded to an adult female chimpanzee, "Pansy," who was ill. Group members stroked and comforted Pansy until she died. After her death, the chimpanzees showed grief-like distress so similar to that of their "human counterparts" that the researchers concluded: "Chimpanzees show self-awareness, empathy, and cultural variations in many behaviors. Are humans uniquely aware of mortality? We propose that chimpanzees' awareness of death has been underestimated."[37]

In another dramatic example of chimpanzee grief, the distress of sixteen chimpanzees was captured on film following the sudden death by heart failure of a female at the Sanaga-Yong Chimpanzee Rescue Center in Cameroon. The "stunning photograph of what they described as a chimpanzee funeral," published by National Geographic in 2009, was picked up by several news organizations, including ABC News and the UK newspaper *The Telegraph*, which reported the story with the headline: "Chimpanzees' Grief Caught on Camera in Cameroon."[38] In the photograph the chimps, attentive and quiet, stare at their dead companion in a way that seemed funeral-like to the reporters.

Deep and powerful responses to death have been seen across species. But when, as Paul Pettitt asks, "did hominoids or hominins begin to develop a 'sense' of death, and a sense that something may lie behind it?"[39] As we have seen, observers of our closest primate relatives, chimpanzees, describe their physical reactions as anguished: they may call out in distress, cry mournfully, carry the corpse, and gather in a group around the body. This evidence of modern primate awareness of death and of expressions of grief lends support to the idea that hominins, too, may have had some sense of death.

Early Hominin Funerary Rituals

Our hominin forebears would have used funerary cairns and caches to hold the dead. Caching involved deliberately transporting the dead to a place where there was a natural opening, such as a cavern or a cave. La Sima de los Huesos, the "Pit of Bones," is the earliest evidence of hominin mortuary practices, and is dated to earlier than 400,000 years ago. The skeletal fossils of twenty-eight hominin bodies were found at this site in Atapuerca, Spain.[40] Later a central site such as a cairn, rather than a cache, began to be used for remains. By the time cairn coverings (such as a pile of stones) were used, mortuary activity involved both natural and artificial elements.[41] Covered cairns, which resemble a simple burial, represented a more deliberate approach to handling the dead.

By the time there were burials, the inhumation of a corpse involved a three-step process: excavating a grave site, interring the body, and covering the body with the excavated material. In his evaluation of the archaeological record of human mortuary activity, Pettitt writes that "archaeologists often regard burial as the developmental summit of human mortuary ritual."[42] It was at this point in our evolutionary history that death became both real and abstract. Purposeful internment meant that rituals had been developed to deal with the dead body, and burial became the benchmark that differentiates human grief from that expressed by non-human animals.

Burials became evident during the Middle Paleolithic (300,000 to 30,000 years ago) and the Upper Paleolithic (30,000 to 10,000 years ago). The earliest burials discovered so far are at Skhul and Qafzeh in Israel, and date back 120,000 years ago (Skhul), and 92,000 years ago (Qafzeh).[43] By the Upper Paleolithic, burials had become ubiquitous. The change from early intentional burial practices to burial ceremonies meant that the living were able to remember the dead in more lasting ways. A conceptual understanding of death had begun: clearly the hominin brain was capable of thinking in symbolic and abstract patterns. And what might lie beyond death became the foundation for belief in religion.

By the Mid to Upper Paleolithic, burials were becoming formalized. In some sites, artifacts such as grave goods and stone markers were buried with the dead. Tools might be buried with a craftsman and weapons with a warrior, and the body sometimes was painted with red ochre (a natural pigment that was also used in Paleolithic cave paintings). The use of ochre and the inclusion of grave goods are often considered "a diagnostic feature of UP burials."[44] Indeed, burying commemorative objects with the dead became the general practice across continents in marked ceremonial burials.

A striking example of a ceremonial burial is that of a Russian Sunghir boy and girl who had been placed head to head in a narrow grave. The boy, estimated to be about twelve years old, and the girl, thought to be around ten, are wrapped in multiple ropes of ivory ornaments. The ivory decorations are estimated to have taken two thousand hours of work to make. The richness of the children's clothing, other grave goods, and the many thousands of ivory beads suggest that the children had high social status in their group. Carbon dating of this burial site, and the funerary customs of these people, place them in the Upper Paleolithic Period, about twenty thousand years ago.[45] By the late Paleolithic Period, mortuary behaviors "had spread throughout the New World [that is, out of Africa and into Europe and beyond], and an increased cultural and regional variability had developed."[46]

Grief across Human Cultures

For contemporary humans, the many ways that the dead are mourned and memorialized are both similar and often dissimilar across various cultures. Gravesites, monuments, and family archives are among the many ways that human cultures remember, honor, and mourn their dead. Some societies remember their lost loved ones with celebratory anniversary activities and other special rituals. And there are societies that deal with death and grief with unusual customs and traditions.

The Torajans, who live in a remote part of the Indonesian highlands on the Indonesian island of Sulawesi, deal with death and its aftermath in remarkable ways. After a loved one dies, the body is embalmed and clothed. The mummified deceased, fully dressed, is then brought back to the family's home until the elaborate, celebratory funeral takes place, which may be years later. During this time, family members talk to the deceased, bring it food, and give visitors access. On the day of the funeral, villagers and relatives meet for the festivities. Torajan kin, no matter how far away they live, are expected to attend. The funeral begins when the deceased—who's been placed inside a replica of a Torajan house—is carried into the village in a procession that is accompanied by banging drums and crashing cymbals. An adult Torajan who as a child slept next to his dead grandfather for years tells Caitlin Doughty, the author of *From Here to Eternity*, "For us, we are used to it, this kind of thing. This life and death." The Torajan people, Amanda Bennett writes, do not escape grief, nor try to, because death is at the center of their lives.[47]

Every spring, in China, the dead are commemorated in a yearly festival called Qingming, or Tomb-Sweeping Day. On this important day, which has been celebrated for thousands of years, many Chinese families perform the traditional Qingming rituals: they go to their ancestors' gravesites to clean and sweep, replace dead flowers with fresh ones, burn paper "money" or incense, offer prayers, and otherwise use the occasion to honor and respect their relatives and ancestors who have died.[48]

Memorial Day is a national holiday in the United States that is dedicated to the memory of all the men and women who have died while serving in the nation's armed forces. Every year, on the last Monday of May, relatives of the fallen, and often volunteers, place small flags and sometimes flowers at the graves. The day is usually marked, too, with a visit by the U.S. president to the Arlington National Cemetery, where a wreath is laid to commemorate all those who died in battle.[49]

American funereal and mortuary services are governed by local, state, and national regulations, and guided by the religious and ethnic customs of the deceased and their family. After a wake, Catholics may say a Mass, with the coffin placed in front of the church's altar during the service; burial usually follows. According to the Jewish religion, the deceased is buried within twenty-four hours without being embalmed, which is prohibited by the faith. In some religions, a simple memorial service may follow a cremation. There are other faiths that have more ornate traditions and longer services. Eastern Orthodox rituals, for instance, include a wake followed by an open-casket church service. At the conclusion of the service, the congregation may walk in multiple processions: one while leaving the church, another at the cemetery, and yet another a week or so later at the cemetery.[50]

No matter our country, our country of origin, our culture, or our religion, we begin to express our early grief through the structured progression of funerary events. The finale of these after-death services is the burial, or some other type of treatment for the body, which is the last time the bereaved is with the deceased loved one. After we bury or cremate our dead, we feel that the lost loved one is truly gone. We grieve their loss, but we keep them in memory. These are among the evolutionary and physiological differences between us and non-human primates and other animals. And although our human grief is likely much more complex and enduring because our brains are different, the grief we feel after a loved one dies is also, at a deep level, what many creatures experience but can express and feel only within the constraints of what they are.

It seems that almost wherever we look throughout the miraculous pageant of animal evolution, we see evidence of grieving behaviors. As Carl Safina explains so eloquently, "Grief didn't just suddenly appear with the emergence of modern humans. All began their journey in pre-human beings. Our brain's provenance is inseparable from other species' brains in the long cauldron of living time."[51] Grief cuts across the taxa of mammals, reptiles, and birds; it may even cut across time back to creatures, like the mighty, intelligent dinosaurs, that lived hundreds of millions of years ago.[52] Today we observe reptiles that stand by the body of a dead companion, and we wonder about a turtle standing for hours in front of a photo of his dead mate. We ache to learn of the visible distress of an orca whale mother depleting her strength in trying to keep her dead calf afloat; or of chimpanzees emitting anguished calls, trying to groom a deceased companion, or grimacing with the facial "grief-muscles" that Darwin described in humans and in non-human primates. Grief is as ancient as species that lived millennia ago. Whatever mechanism for survival is embedded in human and non-human primates and other animals, grief is part of it.

Even so, when we are stricken with grief after the death of a loved one, these connections may not seem to matter. Our heartbreak feels overwhelming and, however ancient its origins, deeply personal and unique. In our highly evolved human brain, we hold memories of our lost loved one, memories that are joyful, sad, painful, humorous, and loving. To return to where we began, to a hospital setting in the middle of an afternoon, we see a grief that seems as ancient as life itself and yet devastating in its immediacy: a mother at the side of her son, a son who said "Bye, Mom!" when he left home that morning, and died before his mother's shift was over.

TWO

Forms of Grief

Grief takes many forms, Flavia, she said quietly.
—ALAN BRADLEY

Grief—timeless, universal, and complex—can take many different forms. It may be anticipatory, forbidden, or ambiguous. It can be acute or severe and prolonged. And different forms of grief can overlap like intersecting circles in a Venn diagram.

To describe fully what grief is, and how it enfolds and suffuses survivors in different ways, requires explaining what is happening in the brain as well as the body. The profound sorrow of losing a loved one affects the entire self of the bereaved. Whatever form of grief is experienced, it is as distinctive as an individual's DNA: the experience of grief, however universal, is unique to anyone who loses a loved one. The form grief takes will depend on many circumstances, but one of the most important factors is the nature of the survivor's attachment to the person who has died.

The Significance of Attachment

What does it mean to be attached to someone, to be in a relationship with a loved one? The origin of the word "attachment" dates back to the eleventh century when "atachier" was an Old French legal term that meant to fasten. Even before atachier was used, a tenth-century word, "estachier," meant to stake up, fix, or support.[1] Now, centuries later, the meaning of attachment remains relatively the same. It is the term used for relationships that involve "security and caregiving," a definition that can apply to adults and children alike.[2]

Adult attachment is shaped by the early child-parent relationship, and begins when the world and the baby are new to each other. In his book *Home Is Where We Start From*, Donald Winnicott details that what is needed for a secure attachment to develop is the presence of "a good enough mother (or parent)."[3] The mother adapts to the infant's needs while allowing the child to experience some frustration. To be securely attached is to be able to tolerate separation. In this favorable setting, a child can experience real feelings and develop a "true self" rather than a "false self," which is the self that develops to please the parent, and later others.[4] To be securely attached is to develop trust in the world, including the trust that the loved one will return.

How separation is handled by the parent and experienced by the growing child shapes how later separations are managed.

Separations are a pivotal part of many developmental milestones, and they are negotiated in ways that depend on the strength of the child's attachment to parents and caretakers. Ideally the child develops a greater tolerance for separation. As the child's confidence that the parent will return grows, longer periods are tolerated without the child feeling abandoned. John Evelyn's 1664 description of a protective line of holly trees—"a hedge of holly, thieves that would invade, repulses like a growing palisade"—can serve as a metaphor for the ability of the child to use her attachment to tolerate separation.[5] As her attachment strengthens, it "repulses" feelings of loss and abandonment.

The history of a survivor's attachments will influence the form their grief will take. An early secure attachment will grow into a psychological bulwark that can weather challenges to the stability of the self, even the ultimate separation: the death of a loved one.

What follows is an overview of many of the various forms of grief. Within each section the forms of grief are organized alphabetically, rather than by how frequently they occur, to help readers find the information they need most.

Most Common Forms of Grief

Ambiguous grief follows news that a loved one is missing, or is presumed dead. This agonizing situation puts family members into a state of suspension.[6] There can be no mortuary and funerary rites if there is no deceased. And without memorial services for the deceased, the bereaved can't experience the transition between what is real (the last time the bereaved will be with the deceased), and what is symbolic (when the lost loved one is interred and becomes a memory). Not knowing what happened to the loved one creates an enduring sadness, combined with an occasional, agonizing flicker of hope that someday information will surface or the loved one will return.

Another time that ambiguous grief may occur is when a biological mother gives up her baby for adoption—for whatever compelling reasons. In this circumstance, ambiguous grief can overlap with chronic grief (described later) to become an undercurrent of lifelong sorrow.[7] The biological mother's loss stretches into a future of unknown circumstances.

Anticipatory grief was first described in 1944 by Erich Lindemann, and it is felt when a loved one is "under the threat of death." Lindemann's paper is based on survivors of the 1942 Cocoanut Grove fire in Boston, and includes relatives of World War II servicemen, who faced a real possibility of being killed in battle. The level of distress caused by the possibility of the death of a loved one may be intense enough to result in an abstract adjustment to the threat of death. Lindemann described such a response as a defense against the reality of death.[8]

The name "anticipatory grief" continues to be used today when a loved one is under the threat of a terminal disease or a lingering progressive illness, such as Alzheimer's or Parkinson's disease. This is grief in the abstract that prepares the primary relative for death's separation. For some who are anticipatory grievers, thinking about and planning for what must be done after the death can ease some of the pain of the eventual loss.

For others, the situation is more complicated. With the cognitive and physical capabilities of the loved one diminishing, painful thoughts can emerge for the caregiver, such as "When will it be over?" Not only is anticipatory grief felt before the death, but thoughts of anger and guilt can also erupt. Compassion fatigue combined with physical exhaustion may also affect the caregiver. And when repeated resuscitation and life-extending procedures delay the inevitable, dying can turn into a dramatic and prolonged experience for family members, adding to the anguish of anticipatory grief.

Disenfranchised grief is grief that isn't recognized by those in the griever's world: family, friends, culture, religion, or society. Grief can be disenfranchised because of the particular type of death. When Kenneth Doka initially coined the term "disenfranchised grief," it referred to stigmatized losses, like suicide, and to losses that are not generally acknowledged by society to warrant grieving, like a pet's death.[9]

More recently, "disenfranchised" has come to refer to the grief that a sibling experiences after losing a brother or sister.[10] In the case of sibling loss, it is the parents of the child who "own" the grief, while a sibling's grief can be marginalized or ignored. When others extend sympathy to the sibling, they often quickly shift to asking about the parents. Sibling grief becomes a satellite sorrow that orbits around parents' grief. Disenfranchised and marginalized, siblings can endure this form of grief for years, until, for some, they are able to acknowledge and experience their loss.

Normal, resilient grief is uncomplicated. Though still wrenching and painful, gradually—only gradually—the bereaved comes to accept the loss, and adjust to life without the loved one. Freud was the first to compare uncomplicated and what is now called complicated grief in his theoretical essay "Mourning and Melancholia." Years later, Lindemann wrote a profile of normal grief, which, like his work on anticipatory grief, was based on survivors of the Cocoanut Grove fire in Boston and World War II. He described the initial reaction of grief after the death of a loved one as "acute," with definite psychological and physical symptoms. The patients he worked with exhibited shortness of breath, sighing, and muscle fatigue. Their somatic symptoms were accompanied by "intense subjective distress described as tension or mental pain." But it was Lindemann's observation that patients who were able to "accept the discomfort of bereavement" were the ones whose grief followed a normal course of mourning. His record of how normal grief unfolds remains relevant today.[11]

George Bonanno's more recent studies of grief are based on bereaved survivors after the terrorist attack that destroyed the Twin Towers in New York City on September 11, 2001. The results of his research show that normal grief can be experienced even after such a devastating disaster. Bonanno describes the recovery of many survivors in the face of this national tragedy as demonstrating "resilient grief."[12]

Less Common Forms of Grief

In **chronic grief,** a sense of enduring sorrow lingers, sometimes for years or even a lifetime.[13] Some dealing with chronic grief alternate between feeling the faint hum of pain in the background and being overwhelmed by any number of reminders: anniversaries, other deaths, or even simply hearing a song associated with the deceased. For others, chronic grief can overlap with ambiguous grief. When this happens, the ambiguous grief of a missing loved one coexists with chronic sadness because of the invisibility of the loss.

The difference between chronic grief and complicated grief (described later) is the intensity of the emotion and the effects on behavior. Those experiencing chronic grief return to their routines even while realizing that they're not themselves. But for the bereaved whose grief is prolonged and complicated, daily life and its routines shut down; they can't function the way they could before the death. Indeed, a change from the low-level sorrow of chronic grief to anxiety or panic could be a precursor to complicated grief. If this happens, seeking professional help is advisable.

Delayed, postponed, or suppressed grief is a sorrow that is not allowed its full expression.[14] It's as though the deep emotional response to the loss doesn't happen, and the grief isn't processed fully. With postponed or delayed grief, the bereaved shows some grief, but what is expressed doesn't seem to be enough given the significance of the loss.

There are a number of circumstances when the bereaved can't allow their grief to be fully felt. How the death happened, or the survivor's own experience of the death event, may be so intensely disturbing that it overrides a full expression of grief. Litigation related to the death can often postpone grief as the focus shifts from sorrow to legal matters. Or perhaps there is not enough social or familial support for a full expression of grief.

Delayed grief can erupt as a full expression of grief months or even years later. A seemingly random event may cause what seems to be an overly intense reaction: a movie showing a child saying goodbye to a father, for example, might cause profuse crying and distress in someone whose father died some time ago, but who wasn't able to grieve fully at that time.

In other cases of delayed grief, the very temperament of the bereaved may be the cause. The bereaved may be accustomed to pulling a curtain down on emotion, to intellectualizing feelings by "thinking things through." Afraid of being overwhelmed by the death of a loved one, the survivor may not allow enough psychological space for a full experience of grief in the time following the loss.

Forbidden grief is similar to disenfranchised grief, but it is more powerful. Typically the loss is owned and acknowledged by the primary griever, but usually only during the funerary services. When the memorial services are over and the estate issues have been settled, the primary bereaved may disown the death and insist that other family members never again speak of the death or the deceased. This can happen in situations where the death is unexpected or not socially sanctioned, like suicide. When permission to express grief is forbidden, family members, especially children, may experience shame at speaking of the lost loved one. And with time, that shame and avoidance can become internalized and cause its own psychological distress.

Forgotten grief occurs when the death event is not acknowledged by others as significant, or when the bereaved's experience of grief is not valued. The experience of a stillborn birth is an example of an underappreciated loss that is often forgotten by other people.[15] The heartbreak of a stillborn baby doesn't elicit the kind of attention or understanding from others that accompanies the death of a baby who is born and then later dies. The emotional effects of a baby who dies sometime after the birth is almost always felt more quickly and with greater force than the loss of a stillborn baby. And parents of a stillborn baby can experience an anticipatory loss after learning that the fetus has died in utero. One tragedy is not less than the other: a stillborn loss and the death of a newborn or older baby are both devastating. But the words of condolence and the form of grief are different. And the grief from the loss of a stillborn, given the lack of social acknowledgment, may get pushed into a forgotten and less than conscious realm. Like some other forms of grief, then, including chronic grief, forgotten grief can linger.

Among those who may experience forgotten grief are grandparents after the death of a grandchild. Grandparents have been called the "forgotten mourners" in this situation, because they are not primary grievers.[16] But grandparents can experience this sorrow in very powerful and overlapping ways. They grieve the loss of their grandchild—the potential of the life that was lost—and they grieve the loss for their adult child, the parent, as well as often for both parents who have lost their child.

Least Common Forms of Grief

Abbreviated grief is expressed by the bereaved, but the recovery happens within a short time. It may be because a new relationship, like a marriage, happens soon after the loss. Or the family and communities—financial, social, or religious—may offer strong support that helps the bereaved. Abbreviated grief is different from normal, resilient grief because of its abruptness. John Archer describes several cross-cultural influences on grieving that may affect the typical length of time spent in mourning. In an ancient mid-American Mayan culture and a modern U.S. West Coast Samoan one, for example, the communities allot only a brief time for mourning the death of a loved one.[17]

Absent grief is an unusual form of grief first described by Helene Deutsch, a psychoanalyst, in 1937.[18] Absent grief is precisely that. No sorrow is felt or expressed by the survivor after the death of a loved one. Grief has gone missing. Absent grief happens when an early attachment has been so fragile that the child wasn't able to develop a tolerance for separation. Perhaps the mother or primary caregiver wasn't there enough for the child to trust that she would return. Or even if the mother was there physically with the child, she wasn't there emotionally because of her own grief, depression, or other difficulty.

When Deutsch wrote about absent grief, attachment as a psychological entity wasn't in the clinical canon. In her paper on this form of grief, she described several cases in patients whose sense of self ("ego strength") wasn't strong enough to face the effects of a loved one's death. For them, to acknowledge the loss would be to experience an overwhelming, fragmenting pain of separation. To protect themselves, the grief had to be completely repressed.

Complicated (or prolonged) grief is relentless. The bereaved continues to yearn for the deceased, and is unable to accept his or her death. Thoughts of the deceased keep intruding, leading to an intense preoccupation with the lost loved one. Often the bereaved will try to keep a close connection to the deceased by hanging on to remnants of the time before the death. For example, a child's room may be kept as it was before the child's death, with toys and books unmoved. Or a widower who cannot bear to part with his wife's clothes might leave them where she put them in her bureau and her closet. Complicated grief can last for years. It occurs in about 7 to 10 percent of the bereaved in the United States, and 2 to 3 percent worldwide of those grieving.[19] In 2015, there were over 2.7 million deaths in the United States (of over 55 million deaths worldwide), which means that between 189,000 and 270,000 bereaved individuals that year may have experienced complicated grief.[20]

Complicated grief can occur due to a number of factors, including a history of insecure attachments, increased stress, and a lack of family and social supports, as well as a vulnerability to depression and anxiety. The symptoms and intensity of complicated grief can affect the ability of the bereaved to participate in ordinary activities and plan for the future, as well as cause feelings of being alone and isolated. For those experiencing complicated grief, professional treatment is advised.[21]

Empathic grief is often felt at a funeral, or when extending condolences to someone who has recently lost a loved one.[22] Reading an obituary, a newspaper article, or a novel, or seeing a movie, can trigger empathic grief, especially if there is some overlap with the person's own experiences with grief. Even if the loss happened years earlier, hearing of another death can, without the person realizing it consciously, reawaken grief from those previous losses. Empathic grief almost always subsides within a reasonable time.

In **exaggerated or excessive grief,** the bereaved continues to feel agitated and unsettled long after many others would have felt some relief. Pacing and crying can accompany self-recriminating thoughts, such as the notion that if only he or she had done something different, the loved one wouldn't have died. The focus shifts, usually without conscious awareness, away from the deceased to the bereaved and feelings of not having done enough. To respond to a death over which one has no control with thoughts of responsibility for the death—as unrealistic as this may be—can return a sense of control. If exaggerated grief persists and develops into frequent episodes of anxiety, these symptoms can be helped with treatment. The overwhelming distress from the loss can also be addressed in individual or group therapy.[23]

Masked grief is an unusual form of grief that is somatic, involving the body of the bereaved in some way. The bereaved does not experience the loss emotionally; rather, the grief is displaced as physical symptoms. These symptoms may even add up to a "facsimile illness," which becomes a way to identify with the symptoms that the deceased experienced either well in the past, or just before death. Usually the bereaved are unaware that their somatic symptoms are connected with the death of the loved one, and that these symptoms are concealing their grief.[24]

Sometimes the death of a public figure will elicit an intense reaction of grief in the community at large. Or when an entire nation is struck by mass distress and shock, citizens will demonstrate a shared public grief. This expression of **communal grief** in the public can also lead to **reawakened grief,** as earlier losses that individuals in the community have experienced come back into focus. If the famous person resembles someone familiar (a father, a wife, a daughter), this can add to the poignancy and grief.[25]

Shadow grief doesn't fully surface. It often happens after an early miscarriage, when a woman is faced with the loss of her embryo (a pregnancy until six weeks of development). The loss of such a small, fragile embryo may seem less like a death than a beginning that doesn't materialize. It's a life unrealized. A woman's grief, and sometimes her husband's, stays shadowed because this is a loss that remains private, unrecognized by others.[26]

Traumatic grief is multilayered and linked to the details of the death, which can intensify the grief experience. Jacqueline Kennedy's horrific experience sitting next to her husband, President John F. Kennedy, when he was shot is an example of direct traumatic grief, because she was right there, experiencing the paralyzing shock and harrowing distress of her husband's murder.

Trauma can also be imagined, for example when in the case of a suicide, a homicide, or an accidental death the loved ones learn details of what happened. The intensity of grief experienced in such cases can be affected by how much detail is shared. All parents would of course be devastated to learn that their teenage son had been killed in an auto accident. But to hear that a tractor-trailer truck crossed a double-yellow line and smashed head-on into their son's small car would probably make it harder to cope with the loss, because the parents could picture the scene and imagine the pain and distress of their loved one before death.

The severity of a real or imagined trauma in survivors can be accompanied by post-traumatic stress disorder (PTSD) responses.[27] PTSD and grief have been referred to as "traumatic grief." It is important, then, for bereaved survivors who may be experiencing trauma to realize that these conditions may coexist, and for clinicians to assess the severity of any trauma, the possibility and severity of PTSD, and whether complicated grief may be developing.

When the traumatic loss occurs in the course of military service, more layers are added to the experience of traumatic grief. For some, especially those in active war zones, serving in the military involves being confronted with imminent death. And when a fellow servicemember dies in combat, grief can weigh even more heavily given the extreme stresses of combat as well as survivor guilt. Survivor guilt can result from the personal, direct experience of being unable to save a comrade or it can happen as a generalized feeling of having survived the war zone, unlike others who did not. Military grief in this situation can be catastrophic, and accompanied by PTSD. Veterans of wars and their families experience grief that is particular and unique to the members of the military and their families. Government hospitals and staff are trained to treat this specific kind of trauma and loss.

The forms of grief described here have been culled from classical and contemporary papers that are based on the experiences of clinicians and evidence found by researchers. With the exception of complicated grief—which is labeled "persistent complex bereavement disorder" and considered one of the "conditions for further study" in the fifth edition of *Diagnostic and Statistical Manual of Mental Disorders* (DSM-5)—these forms do not belong to any existing diagnostic category, and are meant to be explanatory terms rather than templates into which any one grief can fit. Psychiatrists, psychologists, social workers, and those from other disciplines are familiar with some of the most common forms of grief, such as anticipatory, resilient, and complicated. Clinicians also may use different terms for some of the forms of grief described here.[28]

The consensus among clinicians is that most individuals who are bereaved can experience grief on their own without professional treatment. J. William Worden notes that when a loved one is lost, remembering them in various ways, such as by making memory books or simply reminiscing, can both keep a meaningful connection and ease the pain of the loss.[29] It is when grief doesn't change, when the pain and symptoms linger or even increase, that professional intervention can help.

And just as grief is different for everyone who grieves, each loss will be experienced with its own weight, its own sorrow. As Julian Barnes, quoting E. M. Forster, wrote of his own grief, "One death may explain itself, but it throws no light upon another."[30]

Grieving as a Nation

Ceremonial grief, which is marked by funereal pomp and large memorial gatherings, can happen when there is a death of a national personage. After national catastrophes, assassinations. or deaths of national heroes, it is important for the nation to grieve in order to heal. When President John F. Kennedy was memorialized, the images that were shared of the ceremony and of others' grief became a way for the public to join in the experience and grieve as well. Jacqueline's tears hidden behind a black veil, the salute of their toddler son during the funeral parade, and the tender way their daughter held her mother's hand—these images and more were important for the public experience of grieving.

Such public expressions of grief, accompanied by solemn ceremonial rituals, harken back to what was practiced, even expected, in Victorian and other times when grief was expressed by a nation or even an individual. In the mid-nineteenth century, Americans observed conventions that governed how they grieved. Mourning rituals were formal; there were rules of etiquette and specific behaviors that were published and followed, according to class and other social structures. For many, black clothing was a necessary external expression of grief. A Victorian woman in full mourning, for example, would typically wear all black clothing—including a "weeping veil" made of black crepe cloth—for a year and a day after the death.

For nineteenth-century Britons, the death of Queen Victoria's husband, Prince Albert, in 1861 set the tone and the customs for mourning. These accoutrements of death continued throughout the Victorian era, though later in this period, they came to be viewed as part of an obsession with death and its trappings.

In America it was different. The staggering rate of death in the North and the South during the Civil War, from 1861 to 1865, forced significant changes to how grief was expressed. Around 750,000 people died in the conflict, from both the North and the South, or about 2.4 percent of the population.[31] To give a sense of the magnitude of that loss, in 2016, the United States had 318,892,103 residents.[32] If the same 2.4 percent of lives had been lost in 2016, that would have amounted to an astonishing 7,653,410 deaths.

Given these enormous losses on both sides, it was impossible for Americans to grieve as they had before: there was no longer time or money to mourn in the usual ways. And in the wars that followed the Civil War—World War I, World War II, the Korean War, and the Vietnam War—public grief continued to fade from view. For many, the custom of wearing black was no longer required, at least not for as long. And after World War I, more subtle showings of grief, like the "remembrance poppy," were displayed to honor the military personnel who died. This symbol, adopted by the American Legion to commemorate American soldiers who died in World War I, was inspired by the poem "In Flanders Fields," by John McCrae, a Canadian physician. It is still used as a remembrance symbol in many countries, and remembrance poppy pins are still worn on the lapel as a subtle but visible reminder of loss.[33]

Later in the twentieth century, bereavement in the United States became more solitary, and devoting less time for mourning came to be seen as reasonable. The trend has continued today so that now, many years later, grief not only has become private, but is also often hidden, with little time allotted for coping with a loved one's death. As the theater artist Kaneza Schaal noted when discussing her performance piece based on the death of her father: "In the U.S., somebody dies and people take a day off work and go back to work the next day."[34]

Freud and Grief

Sigmund Freud's classic essay "Mourning and Melancholia" was the first effort to describe grief from a theoretical perspective.[35] His treatise was brief and not based on empirical evidence, but it was an important precursor to descriptions of normal and complicated grief. Freud wrote that grief could either unfold as purposeful mourning (normal), or remain internalized as melancholia (complicated grief). Mourning, he noted, was an active way of working through grief. This process—he called it "grief work"—involved a gradual detachment from the deceased, and allowed the survivor to process the loss in a reasonable time. But to experience melancholia was to be trapped in an immovable connection with the deceased, unable to accept the loss. It prevented the survivor from letting go. Darian Leader, a British psychoanalyst, put it this way: "In mourning, we grieve the dead; in melancholia, we die with them."[36]

Phase and Stage Theories of Grief

After Freud, grief was studied in the context of attachment theory. John Bowlby, a British psychiatrist, and Donald Winnicott, a British pediatrician who later became a psychoanalyst, were influential figures in attachment. Their interest began in England during World War II, when parents sent their children away to protect them from the bombings in London and other cities. Bowlby's work on attachment and loss in those children and others was later expanded to include theories of attachment in adults.

A number of clinicians later used "phase" and "stage" approaches in their studies of grief. In their study of grief in adults, Colin Murray Parkes and John Bowlby used phase models in England, while in the United States, Erich Lindemann and

A. Beatrix Cobb used both phase and stage approaches to study grief.

It wasn't until Elisabeth Kübler-Ross's first book, *On Death and Dying*, was published, however, that stage theory became embedded into the American consciousness. Her classic book was not about stages of *grief*, but about the stages of *dying* that the terminally ill experienced.[37] Kübler-Ross's definitive book had a significant influence during the early days of palliative and hospice care, and it caused a fundamental shift in medical and societal attitudes toward dying. Stage theory for the terminally ill came to be the gold standard for forecasting and explaining the flux of emotional responses to one's own predicted death.

As a result of Kübler-Ross's stage theory of dying, the idea of five predictable stages of grief for survivors emerged and became as entrenched as her original idea, though there was no actual connection. As many have noted, the central problem with the stage approach to grief is that the bereaved do not experience grief in five predictable stages.[38] And there is no reason to think that they should: the experience of being terminally ill and facing death is profoundly different than that of being bereaved and struggling to face life without the deceased. Death for the dying ends a life. Death for the bereaved begins a new life—without the loved one.

Contemporary Theories of Grief

As it became clear that grief was being straitjacketed into stages, and not allowed its natural flow, clinicians, writers, artists, and many who were bereaved rejected the five-stage theory of grief and offered their own insights.[39] Research conducted during this time gave a needed boost to non-prescriptive approaches to grief. The research of George Bonanno, for instance, showed that survivors of trauma can experience normal, "resilient" grief, even after catastrophes like the attack on the World Trade Center in New York City.[40] And in their dual-process model studies, Margaret Stroebe and Henk Schut describe how grief can oscillate from a "loss orientation" (in which the bereaved manages negative emotions) to a "restoration orientation" (where the bereaved makes the life changes needed to adapt to life without the deceased).[41] Worden's task model, too, which includes four tasks of mourning, is similar to Freud's theory of mourning that involves grief-work. The four tasks help the bereaved to participate in the painful adjustment to the death in the context of his or her own life circumstances.[42]

Today, such phase or stage theories have been replaced by evidence-based research demonstrating that grief is as unique as each person who grieves. Grief, viewed from an all-encompassing perspective, includes a history of attachments, various life-shaping events, and present circumstances. And it is not for relatives, friends, or society to mandate—to dictate—how a person "should" grieve, because there is no one right way. For each person who is bereaved, their grief is their own. Consequently there is no theory or advice that can provide a schedule or a roadmap of how someone should grieve. Grief—unpredictable, changeable, and uncontrollable—needs to be felt, allowed to flow, and expressed.

But how can grief be expressed? Does it have a language? Yes, with its own vocabulary. This language is not willingly learned, yet it is essential to the act of grieving. The survivor has to be able to give voice to words that acknowledge the death of someone they love. How do they, the bereft, the widowed, the bereaved, or the bereaved parent—some of grief's words—learn and use these words, words that replace the language of love once spoken to their beloved?

THREE

Language of the Bereaved

I tell you I could speak again: whatever
returns from oblivion returns
to find a voice.—LOUISE GLÜCK

To participate in the world around us, we turn to our primary instrument: language. Language expresses what we think and what we feel. It's how we communicate our inner reality to others, and how our external reality, as perceived by our senses and constructed by our brain, is conveyed to ourselves.

Looking at the word "language" on this page of words, it is large, or perhaps medium-sized: eight letters, but only two syllables. To pronounce "language" is rather curious: the tongue pushes up to the back of the front teeth as the mouth opens to utter the first syllable. Then taking a downward dip, the tongue lands behind the lower teeth. Then the lips take over, doing a funny squiggle that involves the cheeks and lower face to form the second, rather drawn-out syllable that ends with a whooshing sound. And that's just one English word. There are over six thousand human languages spoken in the world today; many others have become extinct.[1] Whatever the language, though, each has its distinctive words to be spoken, sung, and heard, as well as written and read.

So it is with grief. Grief has its own forms of expression, and can be said to have its own language. In our attempts to describe the elusive emotion we're experiencing—to unravel our tangled thoughts and reconstruct them into a coherent form—we pluck words out of our language and string them together. It isn't easy to find words to express what we feel. The stronger the emotion, the harder it is to find a word to convey it. Sometimes a word seems to be on the tip of the tongue, right there on the edge of language. Or a thought, pushing its way toward utterance, vanishes, as in Osip Mandelstam's poem "The Swallow": "I have forgotten the word I intended to say/and my thought, unembodied, returns to the realm of shadows."[2]

Words, the semantic units of meaning both denotative and connotative, are what we humans use to express ourselves. The denotative meaning of a word is literal and objective, a dictionary definition; the connotative meaning of a word hides beneath the surface, alluding to, implying, and insinuating deeper meanings. Words express our feelings. Words describe our sensory experiences: what we see, what we smell and hear, and what we touch. And we use special words to denote the important people who populate our world: mother, father, child, sibling, partner, spouse.

How many words are there in the English language? Words that are only English, and without including technical, slang, or scientific terms? The *Oxford English Dictionary* estimates more than 250,000 distinctly English words. Another estimate, which includes the assimilation of words other than English, lands on 988,968.[3] Language, with all its words, is at the service of meaning. In Lev Vygotsky's book *Thought and Language*, he writes, "Words play a central part not only in the development of thought but in the historical growth of consciousness as a whole. A word is a microcosm of human consciousness."[4] Language is our lifeline to one another, to our own consciousness, and to our grief.

Grief Words

Bereaved, bereavement, grief, grieving: these words are indelible identifiers. Their meanings are interchangeable, yet also collide. Because depending on how the word is used (as a noun, a verb, an adjective, or an adverb), its meaning varies in subtle ways. "Grief," the emotion experienced after the death of a loved one, is bone-etched sadness that imbues the self of the survivor. "Grieving" is the act of allowing the sorrow of loss, the thickness of its pain, to be felt. "Bereavement" identifies the lost state into which the survivor is thrust in the aftermath of a death. "Bereaved" can be a noun or an adjective. As a noun, it is a placeholder for the person—the bereaved—who has lost a loved one.

As an adjective, it distinguishes the word it precedes. To be described as a bereaved mother is to attach the loss of a child to the meaning of mother.

In whatever form "bereaved" is used—as an adjective that modifies, a noun that labels, or a state of being—the meaning from the fourteenth-century English "bereafian" remains the same today: "to deprive of, take away, seize or rob." Bereft means "to be robbed of."[5] This is precisely what death does. But the bereaved is robbed of two lives: the life of the loved one, and the life that was lived with the loved one.

"Mourning" is a word closer to times past than present. Perhaps as a carryover from the Victorian culture, it is a word associated with people who had time to grieve, albeit in prescribed ways. That a person is "in mourning" doesn't fit in today's world, where grief is considered little more than a fleeting state. Within society's mandate, grief is not permitted much time: a week, a month, maybe six months. The emptiness of what it means to lose a parent, a spouse, or a child is allowed little more than a brief interval of sorrow.

To those who are newly grieving, words that carry the emotional heft of grief may have been known, but as part of a vernacular that belonged to others. Now these survivors, too, are faced with learning words that are part of a painful and unwelcome language, one that will seem unfamiliar and ill-fitting even as they become immersed in the programmed funerary protocols that unfold over the first days and weeks.

Lost Language of a Shared Life

When death breaks off a life, the special language that the survivor and the deceased used with one another—loving, playful, angry, admiring words, as well as the innumerable other ways the two communicated, like touching, gazing, and laughing—all that disappears. As Julian Barnes poignantly describes it after the death of his wife: "You feel sharply the loss of shared vocabulary, of tropes, teases, short cuts, in-jokes, sillinesses, faux rebukes, amatory footnotes—all those obscure references rich in memory but valueless if explained to an outsider."[6]

The immediate aftermath of hearing that a loved one has died can affect the frontal lobe, the part of the brain that is active when we think rationally, when we make plans or decisions. As one of four lobes—the others are the temporal, the parietal, and the occipital lobes—the frontal lobe is one brain area where "top-down processing" occurs. Because we have a frontal lobe, we can think abstract thoughts. We can imagine a future; we can remember the past. It's because of our frontal lobe that, as Shelley wrote, we can "look before and after, we pine for what is not."[7] When that amazing part of the brain is working at its best—when the requisite neurons are firing—we plan, we judge, we imagine, and so much more.

When news of a death reaches the survivor, abstract thinking—so very important to the decision-making orchestrated by the frontal cortex—is overridden by our senses, which drive the brain's "bottom-up processing." These systems process what we are perceiving through sight, sound, smell, taste, and touch. Bottom-up processing hijacks top-down processing, taking over the systems that are conceptually driven. The unleashed cascade of sensory inputs when shock and disbelief occur can make us feel like we are in a movie. As the experience becomes fixed in the emotional memory of the amygdala, where we were, what we were doing, what we were looking at—even small details—may become eidetic, indelible mental images, ones we keep forever and can replay with great fidelity.[8]

Helen Macdonald describes the moment when she answered the phone and heard her mother tell her of her father's sudden death: "listening to my mother and staring at that little ball of reindeer moss on the bookshelf, impossibly light, a buoyant tangle of hard gray stems with sharp, dusty tips and quiet spaces that were air in between them and Mum was saying there was nothing they could do at the hospital, it was his heart, I think." Macdonald remembers her father as he was days before, and thinks: "That was when the old world leaned, whispered farewells and was gone."[9]

In such powerful moments, the survivor can be rendered speechless. Whenever news of a death is heard, life abruptly stops. But it's not just an interruption. Something has broken—a loved one has died—and this is impossible to undo and make right. The nascent griever, pushed into death's vortex, faces an unknown domain: death, with grief as its language.

In this changed world after a death, a bridge to this new vocabulary is created by the family rituals and funerary protocols that envelop and support the griefstruck. As in the first sentence of Jean Rhys's novel *Wide Sargasso Sea*, there exists a common response to tragedy: "They say when trouble comes, close ranks."[10] Across species, threats and death unleash deep, long-established behaviors. For humans, mortuary services are first, followed by funerary and memorial ones. All take place in concert as the survivor, family, and friends gather to observe religious or secular funereal rituals, each of which operates within a unique linguistic framework. The language of funeral rites provides a uniform means of expression, and within this closed, and close, emotional space, the living and the deceased are together for the last time.

After the memorial services end, the metaphorical fences that surrounded and protected the survivor disappear, and the bereaved is left to live life without the loved one. There are no instructions, no manual. Grievers learn the language of grief—its vocabulary, syntax, and idiosyncrasies—by themselves. Words become tailored to each survivor's needs as a new sense of self is formed. A grieving woman, the first to be recognized as a rabbi, told colleagues after a talk she gave, "You described my grieving pattern. Please, use a language to describe it that does not make me 'one of the guys' again."[11]

The vocabulary—including the tenses of verbs—in this new language helps the bereaved to navigate life's new verbal spaces. Grief words have to be repeated and gotten used to. Slips back to the old language happen, especially in the beginning. The wrong word or wrong tense is stumbled over: "My brother will meet you" is a future event that won't happen because the brother of the bereaved has died. Now it is "I'm sorry, but my brother won't be able to meet you. He died several months ago."

Grief language is a continual reminder of the sadness of the loss, and as its words are spoken and heard, it gradually becomes incorporated into the survivor's consciousness. Some words in the vocabulary of grief may include:

Abandoned Ache Adrift Aimless Anxiety Astonish Battered Bereave Broken Churning Crash Cruel Crying Death Denial Desolate Dimensionless Dirge Disaster Discordant Elegy Emptiness Endless Endure Eulogy Fear Fractious Grief Griefless Griefstricken Griefstruck Grotesque Hanging Heart Heartache Heartbreak Howl Inarticulate Isolate Jumpy Keening Knell Lamentation Lost Missing Moan Morose Mourn Oscillation Panic Remember Requiem Restless Sobbing Sorrow Strength Swept Tears Terror Timeless Torture Wail Waves Weakness Weep Wordless Wrench Wring Yearn

When Words Fail

Before words form, and those words tumble into first sentences, is the preverbal time for a child—the stage in human development before there is language. During this time, the very young child's world is a kinesthetic one, experienced with her body.[12]

The preverbal child lives by touching, smelling, hearing, tasting, and looking at the physical world, discovering and learning about it through her senses. And just as an adult uses language to communicate meaning, the preverbal child uses the body. Piaget labeled the first year or two of life as the sensorimotor stage of development, when action and external experience fuse for the child. When a preverbal child holds up a finger and cries, he is telling the mother or other caregiver that the finger hurts. When his mother responds with comforting words and attends to the finger, perhaps describing what happened in a soothing tone and kissing the finger as she calms him, she is putting words to the different sensory impressions—kinesthetic, aural, visual, and emotional—that the child is experiencing. This is one way the preverbal child learns his native language: first there is the body, then an awareness of words, and then the emergence of one's own language skills.

Much like the preverbal child who experiences the world through the body, many who are newly bereaved are thrown back into a world without words. The shock is in the body—felt, not thought. Not being able to speak, being caught in the throes of grief in the body, is a return to what Adam Phillips describes as the child's "inarticulate self, the self before language."[13] As words become available, the anguish often is described through what the body is feeling: "My heart has a hole in it" tells the listener how the grieved body of the survivor is suffering.

To move away from the inarticulate self and to name feelings that describe life without the beloved will take time. In the midst of the death shock, the survivor can find himself wordless. Antoine Leiris describes his agony after learning that terrorists in Paris killed his wife, writing that he "practically lost the ability to speak. . . . I was incapable of thinking."[14]

Why are words unavailable for some after a loved one dies? How can news of a death or trauma short-circuit language so completely that it becomes difficult to speak? The specific place in the brain responsible for the production and expression of speech is called Broca's area, located on the left side of the frontal cortex. If Broca's area is damaged by a cerebrovascular insult such as a stroke, expressive speech and word production will be seriously impaired.

Trauma can also affect Broca's area. In a study using functional magnetic resonance imaging, or fMRI (a measure of blood flow in the brain), trauma victims were asked to recall painful events and when they did, the blood flow in Broca's area dropped significantly.[15] If trauma and shock can have this effect, could newly experienced grief temporarily diminish brain activity in Broca's area? While there is no evidence to date that supports this conjecture, it seems plausible that when words fail the newly bereaved in the aftershock of a death, Broca's area may be affected.

Sometimes, too, the bereaved chooses not to speak, as a way of exercising, as Antoine Leiris puts it, "the right not to talk about it."[16] When death happens, how its shock is first expressed depends on many factors, including the personality of the bereaved. The survivor's history of attachment, the nature of the relationship with the loved one, and whether the survivor was present at the death will also influence initial reactions to the death event. For some, there can be a glazed silence, or almost-silence with only a few words spoken. For others, howls and wails demonstrate how primal first grief can be.

According to one of Winston Churchill's secretaries, when his daughter Marigold, who was almost three, died, his wife, Clementine, "screamed like an animal undergoing torture."[17] The Churchill children, Marigold and her siblings, had been away at the shore with their nanny when the three-year-old girl had become sick. By the time the Churchills, away in different places, were finally summoned, their little girl was gravely ill with an infection that lingered for several weeks and then developed into septicemia, a blood-poisoning condition. With her parents by her side singing a favorite song, "I'm Forever Blowing Bubbles," Marigold died.

After Marigold's death, Clementine retreated behind a stony wall of grief, one that barricaded her feelings from expression. For many years, Clementine could not and would not speak of the little girl. She was riddled with guilt for not having returned home sooner, and endured her guilt-ridden grief in silence. Her form of grief was probably what we now call complicated grief. At the end of the year that Marigold died, Clementine's grief had increased. In her lingering anguish, she alternated between feeling withdrawn and listless, and bursting into "near-hysterical outbursts."

It was only after years of silence that Clementine's sorrow and guilt calmed, and she was able to speak of Marigold to her daughter Mary, who was born after Marigold died. Mary said she "grew up puzzled by the identity of the little girl whose framed picture stood on her mother's dressing table." The girl in the photo had never been spoken of in Mary's presence.

To use words—to speak—is to give up what was kept to the self. What is private and without words stays contained in the solitary world of the grief-stricken. To voice words is to reveal protected knowledge and give what has been silent a public hearing. Not to voice grief—continuing not to speak—can keep the reality of the loved one's death out of conscious reach. Voicing makes what was private, real.[18]

Naming Grief

To name feelings, to give words—either spoken or written—to emotions that are overwhelming, controls the power of those emotions, harnessing them and lessening their intensity. Using words to express what one is feeling is a way of regulating what is being experienced, and has an effect on both the brain and the body. In the brain, neuronal activity changes from the amygdala—the brain's fear center—to the right prefrontal cortex, an area of planning and behavioral control.[19] Physically, the heart rate slows, and the voice drops in pitch.[20]

In the initial aftermath of a loved one's death, the newly bereaved is forced to begin learning the vocabulary of grief. The effects of the death are usually the first things to be described in this new language: the loved one is gone, life with the loved one is gone, and the person the bereaved was, with the loved one, is gone.

But how to learn the more intimate, meaningful words that belong to a language of grief? Rebecca Soffer and Gabrielle Birkner both lost parents when they were young adults. In a *New York Times* piece titled "How to Speak Grief," they put together a glossary of grief words, definitions, and drawings that describe some of the dark, sad, and even comical aspects of grief. They caution that their vocabulary of grief isn't clinical but comes from their own "hard-earned" and painful experiences. The authors write: "Loss is messy, melancholic and often darkly hilarious. It also lingers forever." One word in their wordlist, and its definition, illustrates the bleakly comic nature of what grief is for them: "Anniversary: A type of day that will forever cease to be identified with happy milestones and instead evoke sharply bittersweet memories related to a dead person. Side effects include pensiveness, tears, and inappropriate outbursts." In addition to definitions, the authors use stylized drawings to express the ache and somber humor of their grief glossary. All of this they offer to readers with the advice, "Use it well."[21]

Grief as a Language of Art

Art in any form is a universal language that penetrates the human heart with its *cri de coeur*. The arc of grief has been powerfully expressed in every art form, and by every generation. It is pointless to suggest that one art form—such as architecture, ballet, painting, drama, memoir, the novel, opera, or poetry—moves an audience more, or tells the best story.[22] Each art form is, to its creator, the best way to honor those lost.

Visual art has been associated with hominin burials as long ago as the Upper Paleolithic. Engravings found on the ceilings and walls of the Cussac Cave of the Dordogne River Valley in France, for example, which are estimated to be around twenty-five thousand years old, show an exquisitely rendered array of animals: bison, wooly mammoths, and horses. Along with the human fossils found nearby, they represent a stunning example of art and human burials in Paleolithic Europe.[23] A hand stencil found in the Vilhonneur Cave near the remains of a male fossil, too, offers a hint of a symbolic association between funerary caching practices and the artistic space of caves.[24] One fossil and a hand stencil are not enough to suggest that cave art served as a symbolic way to remember the dead. Yet Paleolithic cave art, and later art in the Pleistocene epoch, point to how hominins used symbolism to capture and represent the real world in which they lived.

In our contemporary world, the literary arts of poetry and prose narrative, particularly memoir, have been used by survivors coming to grips, linguistically, with grief. The visceral power of a grief memoir is its power to thrust the reader directly into the depths of the writer's anguish. An example is C. S. Lewis's classic work *A Grief Observed*. Lewis describes the trajectory of his grief following the death of his wife: "The less I mourn her the nearer I seem to her. An admirable programme. Unfortunately, it can't be carried out. Tonight all the hells of young grief have opened again; the mad words, the bitter resentment, the fluttering in the stomach, the nightmare unreality, the wallowed-in tears. For in grief nothing 'stays put.' . . . Round and round. Everything repeats. Am I going in circles, or dare I hope I am on a spiral?" Lewis disparages the idea that his experience of grief has a sequential order: "One keeps emerging from a phase, but it always recurs."[25]

Lewis's mother died when he was nine years old. John Archer notes that Lewis was sent to another country after she died, and to a school that didn't allow him to speak of her death. Lewis's loss of his mother was a forbidden grief that made his anguish after the death of his wife, in adulthood, all the more intense. In his descriptions, his grief is palpable. He writes how the force of grief, both early and recent, fluctuates as unpredictably as light on a cloudy, windblown day.[26]

Grief can create such yearnings that the bereaved can fantasize that a loved one will return. In *The Long Goodbye: A Memoir of Grief,* Meghan O'Rourke writes about one such experience after her mother's death. She describes a "weeping" cherry tree in bloom on her street, its pink tendrils arcing slowly toward the ground, evoking the shocking idea "that its presence meant my mother was coming back."[27] O'Rourke's imaginings, she knows, are sad, lovely, and impossible.

Soffer and Birkner, in offering their unique perspective on grief, note a similar reaction in those who are grieving. One of the grief terms on their list is "Mourn Mirage." They explain it this way: "The appearance of a stranger who closely resembles someone who died. May result in your following them several blocks just so that you can be near them (not creepy at all)."[28]

Julian Barnes, at once eloquent and insightful, writes of the excoriating grief he feels after the death of his wife. He is reluctant to face a world drained of her. He refers to the bereaved as the "griefstruck," and describes them as though they are self-employed as grief workers. He writes that those who haven't suffered the loss of a loved one haven't crossed into "the tropic of grief."[29]

Memoirs of grief are filled with voices that reflect the authors' suffering. And while some writers use a comedic sensibility to express their pain, their writings have a common theme: the unfamiliarity with what had been familiar. Nothing is as it was. The new space that grievers live in after the death of their beloved is surreal, unreal. What do the grieved say when others ask, "How are you doing?" What words do they use when the deceased is completely ignored in a conversation? The griefstruck no longer inhabit a time or a place where polite discourse works.

Barnes describes dinner with several old friends whom he and his wife had known for many years. Mentioning his wife's name once, then twice, with no acknowledgment of his loss by any of them, provokes his anger. By the third mention, he wants to confront them. He concludes, "Afraid to touch her name, they denied her thrice, and I thought the worse of them for it."[30]

Sheryl Sandberg's husband died suddenly and unexpectedly when they were on vacation. When she returned to work, she was astounded at the lack of recognition of her enormous loss. It was as though colleagues simply slid over it. Why does this happen? Sometimes the problem is that others fear not knowing what to say; they are unfamiliar with the words of grief. But for the bereaved, this vacuum amplifies the isolation and brokenness they feel. Sandberg describes how she eventually gave herself permission to be frank, and to tell others when they asked how she was, "I'm not fine, and it's nice to be able to be honest with you."[31]

A novel, in contrast to a memoir, can use unconventional ways to express grief. Words, sentences, paragraphs, and the form of the text can tease, cajole, distress, and provoke the reader. Max Porter's *Grief Is the Thing with Feathers* does just that. A mother and wife dies in a sudden tragic happenstance—she falls and hits her head. Her husband and their two boys are left stranded in a house, feeling scrambled and disheveled without her. The three are desolate and unmoored. But then the boys and their father discover black feathers in their beds. Confused, they decide to sleep on the floor. During the night, the father hears the doorbell ring twice. When he opens the door—

> There was a crack and a whoosh and I was smacked back, winded, onto the doorstep. The hallway was pitch black and freezing cold and I thought, "What kind of a world is it that I would be robbed in my home tonight?" And then I thought, "Frankly, what does it matter?" I thought, "Please don't wake the boys, they need their sleep. I will give you every penny I own just as long as you don't wake the boys."
>
> I opened my eyes and it was still dark and everything was crackling and rustling.

Feathers.

Then there was a rich smell of decay, a sweet furry stink of just-beyond-edible food, and moss, and leather and yeast.

Feathers between my fingers, in my eyes, in my mouth, beneath me a feathery hammock lifting me up a foot above the tiled floor.

One shiny jet-black eye as big as my face, blinking slowly, in a leathery wrinkled socket, bulging out from a football-sized testicle.

SHHHHHHHHHHHHH.

shhhhhhhh.

And this is what he said:

I won't leave until you don't need me anymore.

Put me down, I said.

Not until you say hello.

Put. Me. Down, I croaked, and my piss warmed the
cradle of his wing.

You're frightened. Just say hello.

Hello.

Say it properly.

I lay back, resigned, and wished my wife wasn't dead. I wished I wasn't lying terrified in a giant bird embrace in my hallway. I wished I hadn't been obsessing about this thing just when the greatest tragedy of my life occurred. These were factual yearnings. It was bitterly wonderful. I had some clarity.

Hello Crow, I said. Good to finally meet you.

*

And he was gone.

For the first time in days, I slept. I dreamt of afternoons in the forest.[32]

The father's "factual yearnings" are thoughts, articulated wishes that have to do with the present, with what is happening in the moment. These yearnings, "bitterly wonderful" because they are concrete, grounded in the possible, are totally different from the emotional longings of wanting his wife not to be dead.

Words, spoken or written, don't work for everyone. Just as there is no one form of grief, there is no singular way to express it; each person has to discover their own best mode of expressing grief. Some will sketch the world they've been thrust into, or even diagram the pain of absence in a journal or on a pad. Freeform colors can also be vivid conveyors of emotion. Jane Brody of the *New York Times* wrote of a coloring-book approach in her column, "Coloring Your Way Through Grief." Brody cites Dr. Deborah S. Derman who, after experiencing many tragedies and deaths, published *Colors of Loss and Healing*, a unique adult coloring book to help the bereaved deal with grief.[33]

The Reawakened Voice

When does a voice surface after the emptiness of death? Consider the concluding stanzas of "The Wild Iris," a poem by Louise Glück:

> You who do not remember
> passage from the other world
> I tell you I could speak again: whatever
> returns from oblivion returns
> to find a voice:

from the center of my life came
a great fountain, deep blue
shadows on azure seawater.[34]

Eventually those experiencing early grief will learn to navigate an unfamiliar world without the loved one. As grief's language becomes more familiar, a powerful voice can emerge from the oblivion of loss, a voice that tells of grief and the ways that life has changed.

As the wild iris "returns" from the oblivion of winter, its colors, the flower's "deep blue shadows," symbolize a change in the bereaved. In grieving the death of the beloved, the survivor is able to live a life that is shaded by the loved one's absence, but is still a life with beauty.

Even after death's destabilization, when nothing is in balance for the grief-stricken, time rolls on. Sorrow pushes the bereaved forward as time turns days and weeks into months. If grief is allowed its course—allowed its takeover of the survivor—a voice will return, one that can narrate the sorrow of loss and the pain of body and mind:

> The brain: "I can't think straight." "Everything seems confused."
> The heart: "I have a hole in me." "I hurt. I ache."
> And the body: "It's hard to move; my limbs feel so heavy."

In this way the powerful voice of the bereaved, once reawakened, will use words never known before not only to describe the sorrow but also, eventually, to memorialize the beauty of the loved one and the life lived together.

When you bump into something that's new, thoughts put into words can be used to help figure it out. Names and definitions, sentences and paragraphs, make clear what wasn't understood. Spoken, written, signed, or sung, words put things in order. The struggle for the bereaved to understand and articulate the meaning of their loved one's death from their own perspective—that of the living—is an important part of grieving. As Paul Pettitt writes, "Death brings the living into the context of the unknown. In doing so it offers the prospect of renewal."[35] And this renewal comes through the new language of grief. What to say, how to say it, and when: this is the struggle that tugs at the heart and challenges the brain of the grief-stricken.

PART II

Physiology of Grief

FOUR

The Grief-Stricken Brain

From the brain and the brain alone arise our pleasures, joys, laughter, and jests, as well as our sorrows, pains, and griefs.
—HIPPOCRATES

Grief strikes the brain the moment news of a death reaches it.

The day my sister was born is impossible to forget because it was two days after news came of the attack on Pearl Harbor. The winter days after her birth were bitterly cold, yet by late March, the new buds on the trees were pale green.

My little sister was gravely ill. Convulsions made her stiffen, and she cried with high-pitched screams until the warm water they placed her in relieved the spasms.

It was still March, two days after spring tried to start, when my mother, an aunt and uncle, and my grandmother took my sister to the hospital wrapped in a bunting, a baby suit zipped up to her chin so only her face showed. They handed her over to the medical staff for care and healing.

"We're so sorry, we did all we could," they told my mother later, shaking their heads.

My mother's "memory pools" would fill with all the words and images of that night, making every detail unforgettable.[1] "No, it's a mistake," she told the doctors. "You don't understand. We brought my baby here for help. Not to die."

My grandmother, a matriarch who spoke broken English, stood beside her disbelieving daughter. Arms crossed against her chest, elbows jutting out, she tried to shield my mother. My uncle reeled with rage, letting out a muffled scream when he heard the doctor's pronouncement. His wife, my mother's sister, crumpled into a chair.

Five went to the hospital—four adults and a baby—but only the four adults left, carrying the baby's clothes folded inside the bunting. My mother said she didn't hold her baby after she died; it wasn't allowed then. Or maybe she just couldn't. Her shock at the death of my sister crushed her. There was no autopsy: my grandmother ruled against it, and my mother gave in, not realizing that this meant she would never know why her baby died. It was from the aunt who was there that many of these family details gradually became known to me over the years.

The family grieved without understanding. All they ever said was, "She died because something was wrong with her brain." Not knowing the cause of a death can make grief more difficult to bear. My mother sat and cried for weeks. And I, a toddler then, became the child who grew up with an internalized mandate to make people smile. I was that little girl who would reach up to pat my mother's wet face as she cried and rocked me, her surviving daughter.

Considering what the family described as my baby sister's symptoms, she may have died all those years ago from cerebral meningitis. At that time, antibiotics were not available, and it probably had been too late to administer sulfur medications. Cerebral meningitis is an infection that affects the protective coverings of the brain, the meninges. There are three meninges, with names that fit how they look, and how they function. The first is the dura mater. Latin for "hard mother," the protective dura mater is a tough fibrous tissue located right under and semi-attached to a person's skull. It loosely encases the brain and the spinal cord in a sac-like covering with a thickness and texture similar to canvas. The arachnoid membrane, the middle meninges, is a thin sheet of connective tissue with delicate fibers that resemble a spider's web. The pia mater, medieval Latin for "tender mother," the innermost of the three meninges, is a fibrous membrane that clings to the brain's surface. The subarachnoid space between the arachnoid and the pia is filled with cerebrospinal fluid. Because of the limited space between the meninges, swelling from an infection in the pia or the arachnoid will increase pressure on the brain. An infection like cerebral meningitis develops rapidly, and if left unchecked it can lead to brain damage or death.[2]

The family, more often than my mother, would sometimes

talk about my baby sister's inconsolable crying, her arched back, and her convulsions. It was her screaming that prompted the family to take her to the hospital that night, but no one expected my sister to die. For my grandmother, her death was reminiscent of another loss, years before. Shortly before World War I my grandmother's first son, Alexander, had died from undetermined causes. She would say that he died because "his little body wasn't strong enough for him to live." Her explanation made her son's death plausible to her, so perhaps eased her grief. To me as a child, however, the story of Alexander's death seemed like a fairy tale.

If mystery obscures the cause of a death, it puzzles the living, and the brain struggles to clarify the unknown into something it can grasp. Knowing the real cause of death won't stop the pain of the loss of a loved one, but it will free grief to be fully felt. No wonder it's said that "knowledge is power."[3] Words crystallize the unknown into something that can be looked at, considered, experienced and understood.

The Shock

Some scream: a visceral, primordial response to death. Others sob. Or the response can be a stunned silence. Incredulity is the survivor's first response to news of a death. The brain strains to make sense of an entirely new reality. Standing in a hospital, listening to a voice on the phone, facing a police officer in a doorway, or reading a written missive, one experiences acute first grief while wrapped in a cloud of shock. Grief isn't yet what it will become later.

From a friend comes this story of a death, which also happened in the early spring:

> In a pickup softball game, a father drops the ball he caught, and falls. His teenage son rushes over and tries to administer CPR, but it's no use. The boy's father has suffered a heart attack and never regains consciousness. For the son, the aftermath of his father's death will carry additional emotional weight. His grief will be complicated by a deepening sense of inadequacy, feelings of guilt, and an enduring sense of failure fueled by the knowledge that he was unable to save his father's life.

Other traumatic circumstances may attend a death. One may witness a killing and be helpless to save the victim. In this most extreme of circumstances, whoever experiences the horror becomes a traumatized survivor. Shock, trauma, survivor guilt, anger, and grief all converge. After her husband was assassinated, Jacqueline Kennedy was described as looking as though she "had just emerged from a kind of war zone where detachment, like anger, can be key to survival." She seemed to be "in a trance."[4] The shots that killed her husband slammed Jackie into a state of traumatized shock. She refused to change her splattered clothes, telling others that she wanted the world to see what had been done to her husband. Then her initial response of shuttered silence changed: she began recounting the harrowing details of her husband's murder. Barbara Leaming writes that Jackie described the horrific details over and over, first to Bobby Kennedy, the president's brother, and later to others who accompanied her from Texas back to Washington.

Survivors of death-threatening situations repeatedly replay the trauma they lived through with tormenting thoughts of "what if?" and "if only." There were many times that Jackie would rehearse what she could have done differently in that defining historical tragedy. Leaming describes the repetitive track that Jackie's thoughts might have taken in the winter of 1963–1964: "If only she had been looking to the right, she told herself, she might have saved her husband. If only she had recognized the sound of the first shot, she could have pulled him down in time."[5] In the years that followed the assassination, the president's widow suffered many episodes of post-traumatic stress, a condition that had not yet been recognized.[6] It wasn't until sixteen years later, in 1980, that PTSD was added to the American Psychiatric Association's *Diagnostic and Statistical Manual of Mental Disorders* and described as a disorder with specific symptoms.[7]

Even a seemingly innocuous event can flood the traumatized PTSD survivor with visual and auditory memories of the original tragedy. The conflux of sensations rapidly causes a fight-or-flight response that mimics the body's initial experience of the trauma. The brain's emergency responder, the amygdala, pulls the traumatized griever back to their experience of the initial events. For many, psychiatric and psychological interventions such as medication and therapy can help to interrupt this process and separate the effects of the trauma from grief.

Lives Overturned

Dr. K, a physician, and his wife had recently moved from one university medical center in Philadelphia to another in Boston. We met at a fall party. Mrs. K told me she breathed more easily now that her husband's clinical work, which had exhausted him and infringed on his time to exercise, had been replaced by administrative work. She said she felt hopeful as she anticipated their life together. They'd have time to go to plays, the museum, all sorts of enriching events. All would be better, she said, smiling.

We ran into each other serendipitously one day and walked into the building, lingering inside, talking. She told me about their recent Caribbean vacation, describing snorkeling and being able to just relax. It was one of the best vacations they'd ever had—"It was glorious." She thought the time away had helped put into perspective the move to Boston, the adjustments, and her husband's new administrative position.

But it was on that vacation when something first seemed to be off, she later told me. A subtle, odd refusal of a muscle to work in the usual way. Her husband showed her how he couldn't keep the sandal on his left foot. It kept falling off. She was puzzled, she said, shaking her head.

By the time we talked again, it was several weeks after her husband's diagnosis of a brain tumor, a glioblastoma. She was devastated. Her husband, she said, despite his training and experience, seemed hopeful. Their adult children were not.

I didn't see either of them again until one morning when I

went into the living room to open a window in our tenth-floor apartment. The day was clear, and the sky was beautiful, blue. I saw both of them below, in the green-lawned courtyard. Dr. K was sitting on the stone wall surrounding the parterres that were filled with flowers. His wife was sitting by him.

I went down and walked over to them. Gracious as ever when he greeted me, he motioned for me to sit beside him on the wall. His wife nodded to me, got up, and walked over to the flowers. Dr. K talked about the weather—he was from the Midwest—then told me about the medications he was taking. His tone of voice was matter-of-fact, as though he were the physician talking to a patient. He said there probably was no hope. He was concerned about his wife. He spoke about the flowers in the beds and how he and his wife hadn't claimed any space in the parterres. We said goodbye.

I heard about his death from a note I received from Dr. K's wife. She wrote to thank me for going out to talk with her husband that day. He'd had a seizure soon after, and fallen in the rose-carpeted hallway outside their apartment. The EMTs had taken him to the hospital where he was on the medical staff.

She told me she couldn't talk to me the last time she'd seen me because my coming down to talk to her husband had overwhelmed her. She said that not many people had interacted with them once the news of his diagnosis had become known. It was as though the news of her husband's illness pushed people away. She told me she would move out soon: she couldn't stay in their apartment, or even another in the building. She couldn't tolerate walking through those hallways every day. It was too painful.

Dr. K's wife left the building and the state, moving close to a daughter who lived north of the city. I heard from her again several months later. Her life, she wrote, was so unlike the one that she and her husband tried to live before he died. To move away was better for her, for she thought of her life in Boston as having "ended when my husband died." She had to move away from the place where she had lost him, and lost her identity as a wife. In her new life, she was known as a "widow," and sometimes introduced that way, which felt new and different, but nothing like the ache she felt for her husband. "Will my grief ever change?" she wrote. "Will I ever go through a day without constantly thinking of my husband?"

Yes, I thought. In all likelihood, yes to both questions.

The Prefrontal—Thinking—Brain

Where in the brain do we *think?* In our prefrontal cortex, the part of the brain often called the "thinking brain."[8] It is here, in the very front of the frontal lobe, that we plan, judge, and make decisions. But we do more than think logically. We think to create: to paint the *Mona Lisa*, to write *A Tale of Two Cities*, to choreograph *Spectre de la Rose*, to compose *Eroica*. And we think to invent the telegraph, iPhones, and space capsules; we think to discover genetic codes. We became more than we were millions

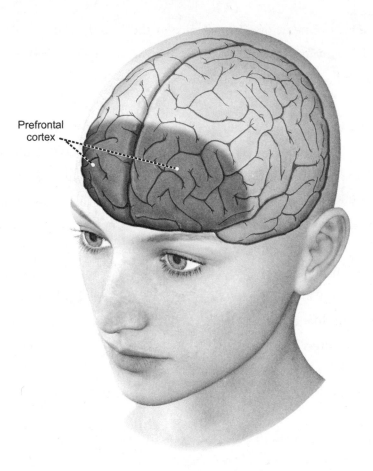

The prefrontal cortex. Illustration courtesy of Steven Moskowitz.

of years ago because the prefrontal cortex kept changing and expanding. It is in large part because of the prefrontal cortex that we think and function in the complex ways we do.

The prefrontal cortex is the largest part of the frontal lobe, and is involved in many kinds of processing. For instance, the orbitofrontal cortex of the prefrontal cortex, which lies directly behind the eyes, has face- and object-recognition neurons that differ from ones in the temporal lobe. These neurons appear to facilitate visual recognition by integrating inputs from sensory information with memory and emotion—helping us to recognize familiar faces or those of people who are significant in our lives.[9] In addition to its planning and decision-making roles, another part of the prefrontal cortex, the ventro-medial region, participates in emotional processing.[10]

The Limbic—Feeling—Brain

Emotions such as fear, delight, anxiety, and anger are the purview of the limbic system, which together with the prefrontal cortex make up the fronto-limbic system. Often called the "feeling brain," the limbic system has an extensive set of structures, many of which are responsible for what we remember, as well as what we experience as excitement, desire, and sorrow.[11]

Emotions can take up temporary, intermittent, or long-term residence in the brain. Performance anxiety is familiar to a speaker who knows that the anxiety will lessen shortly after the talk begins. Math anxiety can pop up when no cell phone is handy and the lunch bill needs a quick calculation to determine the tip. But grief? This emotion is not intermittent in the early months, and it is not temporary. As acute grief after a death turns into grieving, feelings and thoughts continue to spill out in unbidden and unexpected aftershocks. In the matter of human grief—as in survival—the limbic brain is critical to the bereaved.

The key function of the limbic system is to maintain homeostasis of our internal state, to keep us on an even keel physically, emotionally, and cognitively. It does this by coordinating our reflexes and behaviors. There are sensory receptors in the body that send feedback signals, mostly from the viscera (the gut) to brain cells in order to regulate our physical state. These processes keep us alive and are essential in maintaining a state of equilibrium.[12]

The anatomic circuits of the limbic system are extensive. Neuronal networks reach from the cortical, top surface of the brain to very deep within the brain, to the brain stem. The structures that are included in any study of the limbic system will depend on the investigating researcher; no single anatomic definition of the limbic brain has been universally accepted. For many investigators, however, the following four structures are considered the primary elements of the limbic system: the amygdala, the hippocampus, the hypothalamus, and the cingulate cortex.[13]

The Amygdala: Fight, Flight, or Freeze

The amygdala is an ancient area of the brain that is found in humans, in other mammals, and in reptiles.[14] Its origins extend back millions of years, and it is tucked deep within the brain's temporal lobe.[15] One of its functions is to respond to perceived threats by initiating the fight-or-flight reaction that triggers protective and survival behaviors.

Inputs to the bilateral amygdala can come from various senses: an unfamiliar step can be heard, a light flashing in the darkness

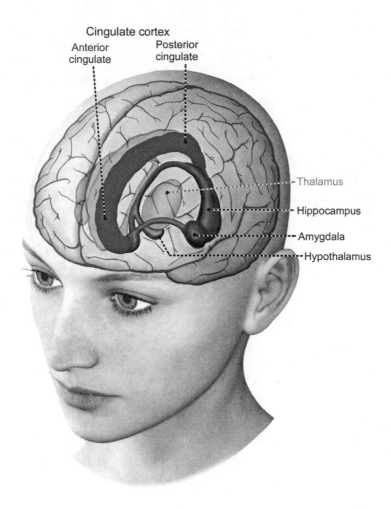

The limbic system—amygdala, hippocampus, hypothalamus, and cingulate cortex. Illustration courtesy of Steven Moskowitz.

can be seen, an acrid odor can be smelled. When humans or other animals perceive a threat, their amygdala instantly responds and goes into a fight-or-flight mode. But if the stimulus is life-threatening and leaves no time to fight or run away, there is another response: to freeze. For humans, to hear of the death of a loved one, what defense can there be? How can you run from death? How can you fight it? There is no escape. This news is unlike any stimulus that can elicit a fight, flight, or freeze response. The survivor's life hasn't been threatened. It's been irreversibly changed.

But this tiny structure doesn't only guard against threats to survival. Research in animals and humans shows that the amygdala plays a role in positive as well as negative emotions via its connections to many brain regions, including the prefrontal cortex and other subcortical regions.[16] How feelings such as ambition, love, and grief are experienced in a situation will depend on the individual. Except for a universal threat to survival or other perceptions of danger, responses of the amygdala are different for everyone.

The Hippocampus: Learning and Memory

When a threat stimulates the amygdala, the hippocampus (Greek for "seahorse," because of its shape) springs into action to access memories and emotions that are associated with that threat.[17] In other words, when sensory stimuli are encountered, the hippocampus and amygdala act in concert: emotion and memory meet, and interpret whether the memory is a positive or negative one. And because the hippocampus has strong connections to the hypothalamus, the regulator of the body's response to threats and stress, the emotional memory signals the hypothalamus to initiate visible changes to the body.

The hippocampus functions in learning and other kinds of memory. It consolidates short-term memory into long-term memory, is important in episodic memory, and is engaged in spatial orientation and navigational memory. London cab drivers are required to memorize the location of more than twenty-five thousand streets in order to qualify for their cabby license. MRIs of the brains of experienced taxi drivers found that they had greater gray matter volume in the posterior hippocampus compared to controls. Their impressive feats of memory support the role of the posterior hippocampus in spatial memory and navigation of the environment. The authors suggest that the hippocampal volume is due to a reorganization of neuronal circuitry rather than the creation of new neurons (neurogenesis).[18] Neurogenesis has been a topic of increasing interest and continuing research in neuroscience. The idea that new neurons are created in the adult human hippocampus is a topic of contentious debate, though it has been shown to occur in the hippocampus of other adult animals.[19] Research has shown, however, that exercise can modify the adult hippocampus, changing its neural structure, function, and organization in response to new experiences.[20]

The important benefits that exercise has on brain function, especially on memory and mood, are an important consideration for the bereaved. Grief can pull those who are mourning a loved one into being less active, and into shying away from being with others, both of which can initiate conditions that could eventually lead to depression and anxiety.

The Hypothalamus: The Thermostat

Just as a thermostat on the wall registers and regulates room temperature, the hypothalamus registers and regulates our response to stress. By integrating information from many pathways in the brain and the body, it keeps the body primed for action and survival.

Located above the hippocampus and the amygdala, the hypothalamus controls changes in heart rate, blood pressure, and the pallor of one's skin. Sweating, goose pimples, and the rate of saliva flow change because of the hypothalamus, which also regulates so-called drive states such as hunger and thirst, sleep and wakefulness, and sex. Given all the different functions the hypothalamus regulates, it seems small: this structure takes up only about 0.3 percent of the weight of the brain.[21]

The hypothalamus also functions as a command center, regulating stress and the resulting release of hormones in the brain and the body. An alarm from the amygdala and hippocampus signals the hypothalamus to release corticotropin-releasing hormone (CRH). This step causes the pituitary to release adrenocorticotropic hormone (ACTH), which in turn signals the adrenal gland cells to release cortisol. The resulting cascade of hormones is called the hypothalamic-pituitary-adrenal axis, or HPA. Cortisol from the adrenal glands shuts down the production of CRH in the hypothalamus. It is this shutdown that puts a stop to the stress response, preventing it from spinning out of control.[22]

The HPA cascade is activated when news of a death first registers in the bereaved and is experienced as a physical shock to the body. The feelings of disbelief and early acute grief will be responded to and regulated by the hypothalamus.

The Cingulate Cortex

Lying deep beneath the cerebral cortex, the outer surface of the brain, is the cingulate cortex. As the largest structure in the limbic system, the cingulate cortex is important in attention and focus, and in processing different emotional and behavioral states. For some psychiatric conditions, such as obsessive-compulsive disorders and post-traumatic stress disorder, studies have shown that the dorsal region of the anterior cingulate cortex is involved. The cingulate is closely connected with the prefrontal cortex, and with the deeper brain regions of the hippocampus, hypothalamus, and amygdala. For some survivors who experience trauma and another's death at the same time, and whose emotional state takes on an obsessive quality, the dorsal cingulate may be implicated.[23] To date, however, no studies have been conducted that investigate the relationship of the cingulate cortex to grief and post-traumatic stress disorder.

Measuring Grief in the Brain

What happens as days become weeks, and weeks become months, following a loved one's death? Are there places in the brain where grief can be measured? Although only a few studies have been undertaken so far, researchers have begun to study grief in the brain. The earliest experiment used fMRI, which measures changes in cerebral blood flow and so indirectly shows variations

TABLE 2. BRAIN STRUCTURES AND THEIR FUNCTION

Brain Structure	Function
Prefrontal lobe	Has key role in planning, judgment, decision-making, and other executive functions involved in abstract thinking
Limbic system	Controls emotional and motivational behaviors; involved in certain types of memory
• *amygdala*	Important in protective and survival behaviors, registers and influences emotional behaviors including positive feelings, and associates emotion and memory
• *hippocampus*	Works with amygdala to process emotion and memory, changes short-term memory into long-term memory, and involved in spatial and navigational memory
• *hypothalamus*	Maintains homeostasis of internal state with links to the peripheral nervous system and the endocrine system via the brain's pituitary gland; regulates the body's response to stress and emotion, including sleep, wake, hunger, sex, and thirst drives
• *cingulate cortex*	Important in attention and focus; involved in both positive (anterior cingulate) and negative (posterior cingulate) feelings
Nucleus accumbens	Involved in motivation, addiction, and pleasure; a significant structure in the "reward system" of the brain

in neuronal activity, to study grief in a group of bereaved women. Women with non-complicated grief viewed photos of the deceased or a stranger that were paired with a grief-related or a neutral word. The posterior cingulate cortex, a limbic structure, showed the most brain activity when the women looked at photos of the deceased paired with grief-related words. There was activity in several other non-limbic regions, including the frontal cortex and the cerebellum, which suggests that because of the extensive neural networks interconnecting the structures of the brain, grief strikes in multiple places.[24] Still it was in the limbic region of the posterior cingulate cortex where grief-related words and photos of the deceased resulted in the most activation. The study was an important beginning for understanding the neurological underpinnings of grief, and how the effects of grief in the brain may be related to how grief affects the rest of the body.[25]

What about the kind of grief that is persistently dark and heavy—that is, complicated grief? As explained in Chapter 2, complicated grief, which afflicts from 7 to 10 percent of the bereaved, can continue for years.[26] In a later study, fMRI was used to compare complicated grief to non-complicated or normal grief in bereaved women. The same paradigm of picture-composites of photos and words that was used in the earlier study was viewed by the women in both groups. In the women with complicated grief, results showed the most activation in a part of the brain called the nucleus accumbens. The women with non-complicated grief did not show activation in the nucleus accumbens.[27]

The nucleus accumbens is part of the brain's reward circuitry. Located in the ventral striatum, it is engaged during behaviors that result in pleasure or satisfaction—for example, drinking when thirsty or eating when hungry. Reinforcing behaviors that activate the reward center of the brain can become addictive.[28]

In complicated grief, certain behaviors that remind the bereaved of the lost loved one can persist and become patterns. For example, keeping items of the deceased the same as they were before the death becomes a reinforcing behavior. Returning over and over to clothing, photographs, and other mementos can strengthen persistent behaviors until they activate the nucleus accumbens, the brain's reward center.

Later neuroimaging studies used structural MRI to measure the brain. In one, researchers found a smaller left hippocampus in bereaved parents in China who had lost their only child during that country's one-child policy. In another, a large population-based study, researchers examined brain structure and cognitive performance in those experiencing complicated grief, those experiencing normal grief, and non-bereaved participants. Their results showed that the group with complicated grief had less gray matter (mostly neurons) and less white matter (mostly axons and dendrites) than the non-bereaved group. Although these findings imply that the severity and experience of grief may be related to smaller total brain volume, an important consideration is whether these structural differences were in place before the loss.[29]

Studying grief in the bereaved using paradigms that can elicit this emotion, and pairing these with neuroimaging, are daunting tasks. Understandably, more research is needed to give us a better understanding of how grief affects the brain.

Silenced Neurons

Other changes occur in the brain after the death of a loved one. Some have to do with the absence of brain activation, now that the loved one is gone. In the fusiform gyrus located in the temporal lobe of the brain, for example, there are face-recognition neurons that fire in response to familiar faces.[30] In times before the loved one died, when, say, the bereaved widow and her spouse were together, these face-selective neurons were activated, along with the parts of the brain that perceive sensory stimuli. Holding one's toddler on a hip while walking on the beach and then recognizing one's spouse in the surf can be processed during the same time, effortlessly and without conscious awareness or deliberation; different regions of the brain respond together as this experience happens.

When the bereaved and her spouse were together at the beach, neurons in many sensory circuits would fire. Imagine the incalculable number of neurons that would have been triggered: visual (seeing her face), auditory (hearing her voice), kinesthetic (touching her arm), and olfactory (smelling suntan lotion). None of these that relate to the loved one will be stimulated any longer. These become the silent neurons of the bereaved. The only such neurons that will fire now will fire in response to photos of the loved one.[31]

In the Aftermath

After the loss of a loved one, and after the shock wears off, the unfamiliar new life gradually becomes familiar. In the slow process of change that Freud called "grief-work," the bereaved becomes accustomed to new places, people, and routines. There are new habits to be learned, new places to discover and navigate, all without the loved one.

Everything that once involved the deceased is different, and when revisited, becomes a new experience. But these experiences do not have the exciting energy that's usually associated with the new. This is grief-stricken newness, and it's disorienting and painful, even as these new cadences of life eventually become more routine. As unacceptable as this may feel in the beginning, these changes will gradually become the new familiar.

The fact of death—that a person is removed from life—may seem simple and stark. A person lives. A person dies. But for the bereft survivor, this change from a living person to a deceased body is complex and profound. No matter the circumstances of a death, whether the death is expected or not, the brain struggles both to process—and to resist—the message of death and the reality of its aftermath.

How do the words take on meaning; how does the brain make sense of this unwelcome message? It begins with auditory vibrations that mechanically displace air pressure and enter the ear, creating neural signals that are processed through the auditory pathways, and, when they reach the cortex—in Heschl's gyrus—are recognized as words. Then for the message to be understood, neuronal firing travels from Heschl's gyrus to Wernicke's area, the critical brain region where comprehension occurs. It is only by taking this anatomical journey that neural signals become the words that are strung into sentences that convey the meaning that a loved one is gone.

Deep in the Grief-Stricken Brain

Much of the brain's processing, for the grief-stricken and for all of us, happens below the level of consciousness. Neuroimaging studies, using EEG (electroencephalograph) recordings, show that brain activity related to voluntary movement actually begins before a person's conscious awareness of the intention to act. As Benjamin Libet explains in his classic paper on "readiness potential": "The onset of cerebral activity clearly preceded by at least several hundred milliseconds the reported time of conscious intention to act."[32] Later studies expanded Libet's findings. Mark Hallett, in his 2016 essay "Physiology of Free Will," describes the results of several of his and other neuroimaging studies, stating, "It appears that there are many things going on in the brain at all times; only one or two thoughts are in consciousness at one time. Much of what the brain does is unconscious."[33] Research continues to make inroads in the brain's involvement in other aspects of unconscious processing. As Stanislas Dehaene notes: "Based on what we now know, virtually all the brain's regions can participate in both conscious and unconscious thought."[34]

What this means for the bereaved is that grief-related thoughts, memories, and emotions can pop up months, even years, after a death. Other feelings that are related to the death, like anger, guilt, anxiety, and regret, can also push into consciousness, seemingly unbidden, triggered by what may seem like the most benign stimulus. A particular song or an aroma of a certain food, for example, can trigger a resurgence of grief from a long-ago loss. Uncomfortable and upsetting as it may be, this is grief plucking the self out of the present and back to the time when the loved one was alive.

Grief resides deep in the human interior. Like fungi that grow deep in the forest floor, connecting the roots of nearby trees, feelings are interconnected and can be experienced whenever something occurs that is reminiscent of, or even tangential to, the experience of being with someone who has died. Like these mysterious fungi, feelings do not always seem predictable or even rational.[35] And just as mushrooms will break through hard surfaces to find light, grief too can push memories of the deceased through to the consciousness of the bereaved when least expected.

Expressing Grief through Generosity

To be born with a dangerous genetic abnormality can mean living under the threat of death. Some parents decide to contribute—in some way—to research in their child's disorder, either while the child is alive or after death. Mr. and Mrs. R, for example, had decided to donate their son's brain to research after his death.

> It was late in the afternoon when I listened to a message from the neuropathologist at the hospital who said he'd gotten a call about a possible brain donation. I called the number and spoke with Mr. R. He said that his son, who had Williams syndrome, had died early that morning from sudden cardiac arrest. Supravalvular aortic stenosis is a cardiac condition that is associated

with Williams syndrome. This heart defect narrows the aorta above its valve, and over time can impede blood flow to the heart.

Mr. R told me that when he and his wife contacted the Williams Syndrome Foundation they learned that a group at a Boston hospital was conducting brain research in Williams syndrome.

When we began to talk, Mr. R told me about his son. He described how appealing his personality was, how he loved music, chuckling when he described how fascinated his son was with vacuum cleaners. As he told me how much he and his wife would miss their only child, he paused and his voice was tearful. He said they hoped what they were doing would help them in grieving their son.

Later, we talked about the details. Mr. R gave me the number and name of the director of the funeral home where their son's body had been taken. The funeral director told me, when we talked, that he'd contacted the diener (an assistant who works with a pathologist using specific protocols to remove brains for donations), who would make the arrangements for us to receive the brain. As Mr. R was saying good-bye, he paused again and said that he hoped this donation would help him and his wife not to feel so devastated by the loss of their only child.

An organ donation after the death of a young or adult child, when parental early grief is acute, is truly magnanimous. In the ancient Egyptian language, "ib" meant heart. When combined with the word "awt," it became the expression "awt-ib," which meant a "wideness of heart." In contemporary English, we would say someone who has awt-ib has a big heart. The bigness of their hearts prompted Mr. and Mrs. R to donate their son's brain to Williams syndrome research, an act of generosity that surely helped them grieve the death of their son.

No matter how it arrives, unexpectedly or after a long illness, the acute grief that follows the news of a death triggers the emergency responses of the body's sympathetic nervous system: increased blood pressure, rapid breathing, feeling faint or queasy, a pounding heart. Any one of these normal responses can occur immediately after losing a loved one. Although they will subside long before the circuitry in the brain has a chance to adapt to a new life, as grief unfolds in the ensuing months, the brain will develop new neuronal connections—creating the new patterns, the new habits, of a life that is happening without the missing loved one.

FIVE

The Broken Heart of Grief

The heart has reasons that reason cannot know.
—BLAISE PASCAL

We say it's with the brain we know, and with the heart we feel.
Our sense of the heart as a vault holding our most tender emotions and most passionate feelings dates back thousands of years
and across a variety of cultures, art forms, and even everyday
sayings. The ancient Egyptians, for example, believed that the
heart was the place of memory, intelligence, and wisdom, and
that it symbolized emotions such as sadness, bravery, and love.
The human heart was so important that when the organs of the
deceased were removed from the body for mummification, the
heart was left undisturbed, protected by a funerary amulet representing rebirth and resurrection that was placed on top of the
mummy. The practice was meant to ensure that the heart would

remain with the deceased even if something untoward happened. A scarab beetle was carved on one side of the amulet, and a prayer from the Book of the Dead was inscribed on the other.[1]

In the performing arts, an actor on stage may pat the left side of her chest to convey that there are no words to express the powerful emotions her character is feeling. And in narrative classical ballets from the nineteenth and twentieth centuries, a dancer's hands, simply crossed over her heart, are enough to tell of a great love. Americans, too, when expressing loyalty to and love of country, perhaps while reciting the Pledge of Allegiance or singing the national anthem, often put their right hands over their hearts.

Human emotions, fragile and intangible, can be overwhelming. But when the word "heart" is used, somehow the load is lighter, more manageable. When said endearingly, "Dear heart"; empathically, "My heart feels for you"; enduringly, "You have my heartfelt sympathy"; or intimately, "Let's have a heart-to-heart talk," including "heart" in the conversation invites the listener to share in the emotion, and to have his own feelings heard.

The Physical Heart

Not exactly on the left side of the body, but to the left of the breastbone, the heart is enclosed within the pericardium, a double-walled sac that surrounds and protects it. The heart has four chambers: the two upper, the atria, are the collection chambers, and the two lower, the ventricles, are the pumping chambers.[2] With its tube-like aorta, its chamber-like atria and ventricles, its lever-like valves, and fuse-like electrical units, the heart is composed of cells specific to form and function: heart cells (or "cardiac cells"). Among the types of heart cells are special muscle cells, called cardiomyocyte cells. It gets more complicated the deeper you dig: cardiomyocyte cells, in turn, are specialized, with some contracting the heart chambers to pump the blood, and others transmitting electrical signals around the heart.[3]

Each heartbeat consists of a contraction, when the muscles of the heart squeeze like a closing fist, and an expansion. Then the heart rests briefly before the next contraction begins. On average the heart beats more than a hundred thousand times in twenty-four hours. The heart pumps blood through arteries that become smaller and smaller as they reach the outer extremities of the body. After delivering its oxygen, the blood flows into veins, returning to the right side of the heart to begin its cycle of circulation again.[4]

In 1628, William Harvey was the first to describe the heart as a pump as well as a muscle, and he recognized that as a pump, the heart circulates the blood. Harvey, a British anatomist, went on to describe the entire circulatory system. Harvey's experimentation and discovery laid out how circulation happened, but not why. It took another hundred years to understand the purpose of circulation: because the right ventricle pumps blood to the lungs, the blood is oxygenated, and once it is carried from the lungs to the left heart, it is circulated to the many vessels that supply the body's metabolic needs.[5]

In her remarkable novel about a heart transplant, *Mend the*

Living, Maylis de Kerangal has the cardiac surgeon recite his ritual homage to William Harvey. He does this as his patient's, the recipient's, damaged heart is lifted from her body and her circulation is entrusted to a machine for two hours. The surgeon thanks the physician who lived centuries ago and first described the hydraulic effect of the heart's pumping action, the activity that ensures the continuity of blood flow with its intricate movements and powerful pulsations.[6]

Can Grief Break a Heart?

Can there be such a thing as a heart breaking from grief? Yes. Acute grief can stun a heart and result in a cardiac condition called "broken-heart syndrome." This syndrome, first described in Japan in 1990, was given the name "takotsubo cardiomyopathy" because of the ballooning of the heart's lower left ventricle.[7] Echocardiograms show that the heart muscle looks "stunned," and its balloon-like shape often resembles a "takotsubo"—a Japanese octopus-trapping pot with a wide bottom and a narrow neck.[8]

When a patient presents to a hospital emergency room with what looks like a heart attack—for example, with shortness of breath and chest pain—it is critical to determine whether the patient is experiencing severe emotional stress. Women between the ages of fifty-eight and seventy-five account for 90 percent of the reports of broken-heart syndrome, which is reversible. Although the heart may also be more at risk of a heart attack or other heart-related health issue soon after a loved one's death, within a month these cardiac risks decline, and within two months most patients are at no greater risk than normal of a heart problem.[9]

Sandeep Jauhar, in his book *Heart: A History*, describes a patient whose husband had died a week earlier. After looking at a photo of her husband, she cried and began experiencing chest pains and other symptoms that suggested congestive heart failure. Her ultrasound showed that "her heart had weakened to less than half its normal function," yet other tests were within a normal range. After just two weeks, however, her symptoms had disappeared and the results of her ultrasound were normal.[10]

The exact cause of broken-heart syndrome has not been identified. The current thinking is that the body's physiological reactions to extreme emotion—grief, fear, even elation—can trigger a sudden surge of stress hormones like adrenaline, which prevent the left ventricle from contracting properly. An influx of adrenaline, in particular, can also speed up the heart, stunning it and causing damage to heart cells. In fact, the abnormal shape of a heart affected by broken-heart syndrome may be due to areas with higher and lower densities of adrenaline receptors. (It seems that intense happiness can also affect the shape of the heart, but differently than grief. When the ballooning of the heart is caused by intense happiness, the enlarged area is in the mid-section.[11])

Broken-heart syndrome has been slow to become recognized in the United States, perhaps because research to date suggests that only 5 percent of women evaluated for a heart attack were suffering from it.[12] But an international team of experts has published a report designed to help clinicians diagnose it when they suspect it. Doctors are advised to take a multidisciplinary approach, using an assessment of neurological, psychiatric, and other symptoms, along with diagnostic guidelines, to help them understand and treat this acute but reversible cardiac condition.[13] For most people, recovery is rapid without any long-term damage to their heart. Nevertheless, for those who have a history of cardiac problems, it may be important to be evaluated for stress-induced cardiomyopathy during the period of acute physical and emotional stress after a death. Of people sixty-five years or older, 45 percent of women and 15 percent of men will lose a spouse, and for those who have an existing cardiac condition, especially a history of one or more heart attacks, the risk of another heart attack the day after their loved one dies increases twenty-one fold. (This risk factor declines over the subsequent weeks and months.)[14] It's true that some hearts hurt so much with grief that they're vulnerable to breaking, even if it's also true that the hearts of most bereaved survivors can withstand a loss and the ensuing grief.

A Writer's Plea

The book *Cry, Heart, But Never Break,* by Glenn Ringtved, is a plea to the bereaved not to let the death of a loved one break their hearts. Ringtved wrote this children's book for his two sons and two daughters after the death of their beloved grandmother, who lived with them.

In the illustrated book, the character Death does not, as drawn, seem especially terrible or scary. Death is robed and hooded in black, but his head is not a skull as it's usually depicted; instead he has a sad and regretful, if not exactly friendly, face. Ringtved's picture-book story centers on a plan the children hatch to trick Death, who arrives one night to take their grandmother, lying sick in her bed upstairs. Death knocks on the front door and the children let him in. Death politely leaves his scythe outside, leans it against the front of the house, while he accepts the children's invitation to sit at the kitchen table, rest, and have a cup of coffee. They believe that Death can only do his work in the nighttime, so they devise a strategy to keep refilling Death's coffee cup until the night is over. The four children imagine that when morning comes, their black-hooded but gentle guest will have to leave their house without going upstairs to take their grandmother.

Morning comes, but it doesn't change Death's purpose. One child, about to pour more coffee into the emptied cup, is stopped. Death puts his hand over the cup and shakes his head no. But before he goes upstairs, he tells the children a story about how sorrow, delight, grief, and joy find each other. And how these feelings—sorrow and delight, and grief and joy—cannot live without each other. Not quite sure of the meaning, the children feel that the story Death tells them is right: without sorrow there is no delight, and without grief, there is no joy.

The grandchildren slowly climb the carpeted stairs to the second floor and tiptoe into their grandmother's room. They see that she has died. Death, who is standing by the window, looks at them and quietly says, "Cry, Heart, but never break."[15]

A Heart That Didn't Break

During early grief, when the loss feels acute and impossible to bear, some newly grief-stricken experience a cardiac event that doesn't break their heart, but threatens it.

Ms. S came to see me several months after her husband died. His diagnosis had been congestive heart disease, a condition that had steadily worsened. After repeated alarms and hospitalizations, his congested heart had faltered to the point where it finally stopped.

Sometime after the funeral, Ms. S experienced several fainting episodes that seemed to have no apparent cause. Though nothing exactly like this had happened to her before the death of her husband, she told me she thought there had been "previews" in the previous two years. One had happened several days before her mother died, which was more than a year before her husband's death. She was at the hospital with her mother, whose condition had become critical. She felt "a whoosh of black" rush over her, but the feeling quickly dissipated, and she didn't faint. A similar incident occurred months later, when she was with her husband during one of his final hospitalizations. Ms. S had been his caretaker, calling numerous times for an ambulance when her husband was in severe medical distress. Again when she found herself at a hospital with a loved one, this time her husband, she felt lightheaded and dizzy. As before, the symptoms subsided, and she thought no more of it. Looking back, it seemed that these symptoms were the first salvos fired across the bow of her bereaved heart.

As a new widow, Ms. S experienced several fainting episodes. After a near miss when her head grazed a piece of furniture as she fell, she sought medical help. The cardiologist recommended a heart catheter ablation. He explained that her heart had probably been enduring irregular, rapid heartbeats that caused her fainting episodes, and he described the catheterization procedure. Radiofrequency catheters are used to destroy the heart cells in the small area responsible for the irregular heartbeats. By destroying the heart tissue that is disrupting the regular heartbeats, the procedure restores the normal rhythm of the heart.[16]

After she learned what was involved in the heart ablation procedure, Ms. S agreed to it. She watched the procedure on the monitor above her as it was taking place.

The heart ablation worked: Ms. S experienced no more fainting episodes. But her grief at the loss of her husband increased. For months she ached and often despaired, as she began enduring living a life without her beloved husband. But something else lurked beneath her widow's grief. Memories of another death, one that had happened when she was a teenager, began to surface and become conscious. It was the death of her father. Ms. S's courage to endure and live her widow's grief was helping her to reappropriate the buried memories and feelings of the loss of her father, who had suffered a violent death. The suddenness and shock of it had unraveled the family, so much so that the death had to be relegated to a forbidden place where no one could talk about it. At the time, too, Ms. S, young and fatherless, could tolerate remembering it only in a starkly intellectual way: all that she would say was that her father had died when she was a teenager. Now, many decades later, as a widow, she was finally ready to face the repressed death of her father and begin to grieve him.

In her therapy, Ms. S and I talked about the different forms of grief that she had experienced: grief following the death of her mother; anticipatory grief as she cared for her husband during his illness; the grief she experienced after he died; and forbidden grief that changed when she was finally able to grieve her father's death. Her griefs would surge, ebb, and then cycle again. Finally, they became quiescent, enough so that on a recent anniversary of her husband's death, she felt "a calmness" on that day that she hadn't felt before.

Dual Deaths

What causes the weakened hearts of loved ones to break together, at almost the same time? How the emotional heart can affect the biological heart so deeply remains a mystery.

I was living across the country when I heard that my maternal aunt, who lived on the East Coast, was in critical condition. My aunt's congestive heart condition had worsened; she was in the hospital and not expected to live much longer.

Years earlier, my grandfather, my aunt's father, had been hospitalized with congestive heart failure at the same hospital. He had died less than twenty-four hours after I had visited him and kissed him goodbye for the last time. My grandfather was stoic, learned and unsmiling, and he and my grandmother occasionally took me on walks. He would walk briskly, still healthy then but still unsmiling. Those walks were also silent, like meals at the dinner table with him. With my grandmother, walks were different, a story-telling time. She'd tell about her life growing up in another country, chuckling or tearful. The farmhouse that her parents lived in was where her toddler sister had died on a bench in the kitchen. It was soon after her sister's death that my grandmother left and came to America.

Occasionally, we three would walk to my aunt and uncle's house, which was painted a very dark brown. I'd be the first to march across the porch into the dimly lit entryway. In the doorway to the apartment, my aunt would greet us, kissing her father first.

Our visits were short. They began and ended in the kitchen, a yellow room with dark yellow wainscoting below yellow-wall-papered walls. The room was like my grandparents' kitchen, but my aunt's didn't have a coal stove to heat it; rather, a white porcelain stove that burned oil from a big glass jug behind it kept us warm. It was in the kitchen where every life event was dramatized, argued over, and resolved.

My aunt and uncle often disagreed when we visited them or they visited us. That's what made the visits short: a disagreement often ended up in an argument. But they laughed a lot, too. My aunt and uncle always seemed to be together.

Now, with my aunt hospitalized, my uncle visited her every

day, staying as long as he could. It was clear to the family how upset he was about his wife. Looking back at that time, he was in the throes of anticipatory grief.

After a week, my aunt began to slip in and out of consciousness. One morning, when my cousin, who lived above her parents, was getting ready to go to the hospital for her daily visit, she heard a thudding crash below. She rushed downstairs and found her father collapsed on the kitchen floor.

When the emergency team arrived, she told them her father had suffered a previous heart attack, and asked if this was another. The drivers said they weren't sure—she would have to talk to the doctors at the hospital. In the emergency room, the doctor who came over to her suggested that they go into a private room to talk. He told her how sorry he was to tell her what happened. Her father had died before he reached the hospital, probably in the ambulance. He'd suffered another heart attack, this one fatal.

She told me she couldn't understand what the doctor was telling her. Her father? Dead? She described feeling as though the doctor's words hung in the air. What she was hearing didn't make sense. It was her mother who was in the hospital—*she* was the one they were expecting to die. How could this happen to her father? How could her father's heart give out, how could he die?

She looked at the doctor, feeling paralyzed. Then she realized that her brother and sister, whom she had called before leaving, would soon arrive and she would have to tell them that it was not their mother who had died, but their father. This unexpected news would be crushing. Everything had been turned upside down. Then she realized something else. What if their mother regained consciousness and asked for their father? What would they say to her?

Her siblings arrived at the hospital together and came into the room where she was waiting. With her voice breaking, she told them their father had just died. They looked at her with disbelief. Had she gotten mixed up, was she okay? She began to speak slowly and to stop between words. No, she wasn't mixed up, and no, she wasn't okay. It was their father, not their mother, who had died. They talked and cried together.

Finally, they went up to the fourth floor to see their mother and tell her the terrible news. But when they got off the elevator, the cardiologist drew them into a conference room. He said he had heard about their father, and told them how sorry he was. He then said he was so very sorry to have to tell them that their mother had died just a short while earlier.

At the funeral mass for my aunt and uncle, everyone kept looking at the two coffins placed side by side in front of the altar. Cloths of dark violet, the liturgical color for requiem masses, covered each coffin. The family knew they'd died in separate places, he alone in the kitchen and she alone in her hospital room—but together in time. Another aunt, smiling and crying in the church, was heard to whisper: "It's like Romeo and Juliet—they're still together."

Had my uncle's anticipatory grief been powerful enough to break his already weakened heart? He and his wife, despite their heated arguments, shared an intense love that kept them inextricably bound. My uncle's visits to his wife, whose heart was failing, had further stressed his own heart, which stopped before he had to face life without her.

The blow of that first death, their father's, would magnify my cousins' grief. It would take more time, more sorrow, and more adjustment to grieve both parents. For my cousin, the daughter who found her father on the kitchen floor, it would be harder. Her doubled grief was compounded by the trauma of discovery.

My aunt and uncle, who wrangled and loved so much, were together in death. In the words of Shakespeare's romantic tragedy, written centuries ago:

> Romeo, there dead, was husband to that Juliet;
> And she, there dead, that Romeo's faithful wife.[17]

A Gifted Exchange

When a healthy heart from someone who dies is donated to someone whose heart is struggling to beat, this gift, this exchange, can mean that someone will live because someone died. In Maylis de Kerangal's novel *Mend the Living*, a young man in his twenties falls asleep at the wheel of his van after hours of surfing with his friends in very cold water. In the crash, his brain is severely injured. His parents are persuaded to donate his organs, including his vigorous heart, after his death.

The heart recipient is Claire, a woman in her early forties. She is a mother with sons, and her damaged heart hasn't been working well for some time. Because it won't last much longer, she's been hospitalized. Although the surfer's blood type is relatively rare, it's a match for Claire's, and serendipitously, in addition to their immune system compatibility, the heart's physical structure also matches hers.[18]

The surgical team prepare Claire, the mother, for the transplant, making her body ready to receive the healthy heart. The final work of transplanting the healthy heart into the opened cavity of her chest has begun.

> The surgeons now begin the long task of sewing: they work to reconnect the heart, moving from bottom to top, anchoring it at four points—the recipient's left atrium is sewn to the corresponding part of the left atrium of the donor's heart, same for the right atrium. . . .
>
> After the heart is slowly irrigated, the "electric moment" arrives:
>
> Shock! The heart receives the shot, the whole world stands still above what is now Claire's heart. The organ stirs faintly, two, three spasms, then goes still. . . . Second try. Clear?
>
> —Shock!
>
> The heart contracts, a shudder, then moves with nearly imperceptible tremors, but if you come closer, you can see a faint beating, and bit by bit the organ begins to pump blood through the body, and it takes its place again, then the pulsations become regular, strangely rapid, soon forming a rhythm . . . and yes, it is the first heartbeat that can be heard, the very first heartbeat, the one that signifies a new beginning.[19]

The gifted heartbeats sound steady, ringing out with a healthy "lub-dub" as the heart pumps blood through the circulatory system of another.

De Kerangal's novel is an exquisite dramatization of anguished parents, shocked and disoriented by this sudden and unexpected death, who decide to donate their son's heart to someone who is on the cusp of dying. And it dramatizes the dedication and skill of the surgical team who make possible the wonder of a heart transplant.

In real life, too, generous parents and others can donate an organ of a deceased family member for a transplant, knowing that the gifted organ will help someone live, while also perhaps easing their own grief.

An Enduring Link

In her article "Guided Down the Aisle by Her Father's Heart," Katie Rogers writes about a donated heart that went beyond replacing a damaged one. The story begins violently, when a fifty-three-year-old man, a father, is unexpectedly robbed, then tragically shot in the head. He is kept on life support, and before he dies, his family decide to donate his organs to the Center for Organ Recovery and Education.[20] This organ donation center allows the families of donors to be in touch with the recipients.

In this case, the recipient is a seventy-two-year-old man who has suffered for sixteen years from a congestive heart condition. He is the father of four: three sons and a daughter. After the transplant, he is able to thank the donor's family and establish a relationship with them.

When the daughter of the man who was shot and killed plans to marry, she realizes, sadly, that her father will not be able to walk her down the aisle. It's at this moment that her fiancé has the beautiful, unusual idea that she ask the man who lives with her father's heart to walk her down the aisle on her wedding day. After his own daughter approves this plan, the seventy-two-year-old man agrees.

When the man and the daughter who lost her father meet, he encourages her to grasp his wrist so that she can feel his pulse— the beating of her father's heart as it pumps blood throughout this man's body. Then the bride-to-be puts her right hand up to his chest and touches his left side, to feel her father's heart beating. She will be walked down the aisle by the man who has the heart that was her father's.

A life continues, health and vigor renewed, because a stranger's heart permits it. The magnanimity of these actions, medical and human, is immeasurable. And nearly beyond understanding. Could William Harvey and his colleagues in 1628 have imagined that a heart from one body would be placed into another?

Even today, heart transplants are complicated and risky. The removal and transplant must happen quickly. In transporting the heart from donor to recipient, distance and time are critical factors determining life and death. The development of a device by the Organ Care System has extended the transportation time. Dubbed "heart in a box," the machine pumps warm, oxygenated donor blood through the heart, which continues to beat, and so extends its life by several hours.[21] And even if the travel time is short, for the transplant to work, the hearts have to match in ways that include body size, blood type, and the severity of the recipient's medical condition. All of these limitations mean that more patients wait for a suitable donor heart than hearts are available.

Yet because someone dies and donates their organs to another, others continue to live. The gift of a heart is bought with the life of someone who dies, and is paid for with the generosity of the deceased and their survivors. For the grief-stricken who decide to donate their loved one's heart, their own heartbreak may be eased by this generous gift of life to someone who had no hope without it.

The amazing heart is the only part of the body we can feel moving with its lub-dub beats. The breaking heart, the threatened heart, the struggling heart, the gifted heart—all are part of its mystery. When it finally stops, a life ends, even as the biological and emotional heart of the bereaved continues on.

SIX

The Grieving Body

Our own body possesses a wisdom which we who inhabit the
body lack.—HENRY MILLER

Death tears apart the world of those left behind. Much like the
cataclysm that destroyed the dinosaurs' world, death changes
the very terrain that survivors have to navigate. The newly grief-
stricken are pressed into a nebulous space between a world that
has suddenly disappeared and one that as yet has no structure or
meaning. The news impacts every aspect of the survivor's life and
self. The loved one is gone, and so is the physical attachment to
them. Sorrow floods the body. Disoriented and untethered, the
survivor struggles to regain some balance, some steadiness.

The traditional services that take place after a death—funereal,
religious, and memorial—provide the newly bereaved some res-
pite in the short term. Other related activities, legal and financial,

will need attention as well. Such after-death tasks give the survivor a way to maintain a quasi-attachment for a brief time.

With the demands and formalities completed, the new griever begins a life without the loved one. Grief will become an intimate acquaintance. The first experience is the physical emptiness: the loved one can no longer be embraced. Tears, which can start at any moment, may change into a listlessness that suffuses the spirit as well as the body. What was pleasurable becomes diminished. The sensory world is left without its former beauty. Colors are no longer vibrant and exciting, but pale and dull. Food can become tasteless—a friend once described how, for many months after her mother died, she could not detect any taste in her food. This is the beginning, when sorrow is experienced in the body of the bereaved.

Struggling to Sleep

How do you sleep when you know that as soon as you start to awaken—when you first slip back into consciousness—you have something awful to remember? What can the quality of sleep be when sleep cycles only thinly veil the awareness of that loss?

Your loved one has died. The exhaustion of not sleeping well, or of being in a fitful sleeping state, makes the nights stretch with pain. Meghan O'Rourke writes, "The nights were long and hallucinatory; death seemed present in the room with me, an enemy to have it out with then and there."[1] Other authors, like Oliver Sacks, write similarly of this period of turmoil, "Life experiences can be so charged with emotion that they make an indelible impression on the brain and compel it to repetition."[2] He describes bereavement hallucinations that are engendered by loss and grief: "Losing a parent, a spouse, or a child is losing a part of oneself; and bereavement causes a sudden hole in one's life, a hole which—somehow—must be filled. This presents a cognitive problem and a perceptual one as well as an emotional one, and a painful longing for reality to be otherwise."[3] These are some of the ways that the body longing for the lost loved one tries to hold on to an image, a voice, or any of the body's experiences of the one who has died.

Sleeping, eating, living—everything is different. You feel dismembered. And in many ways you are. You have lost a part of yourself, the person who, because of that relationship, made you a wife, a husband, a mother, a father, or a sibling. That person, who occupied such a large place in your life, is gone.

In time, the brain will accommodate the new reality, as cells no longer fire for the one who can no longer be seen, heard, kissed, or touched. This is the reality that slowly pervades the mind and body of the grief-stricken. The pain gradually diminishes, but the loss will always be felt, leaving the survivor bereft.

Again that word "bereft," which means to be deprived of, to be robbed. Death stole the loved one, leaving a new world, with its new language, with all its dismal newness. As the British psychoanalyst Darian Leader explains, "This is a long and arduous process, and each person must find the form of language that suits them and their concerns best. This can never be predicted in advance."[4]

It is impossible to know how long the most intense period of grief will last. Creating a life without the deceased involves many changes. The survivor has to learn not only to feel in new ways, but also how to navigate responses to condolence-givers. During the thick of grief's sadness, hearing painful, unwelcome words from others—words that tell you that you will get used to the emptiness of your loss—can be hard. Every part of the griever's self protests words that are pinned to the future, because there is no future, there is only now. In the depths of grief, hearing that "things will get better" can hurt. Being told that "you'll be okay" can feel like a betrayal of the deceased. It's as though if you begin to feel better, somehow that means your loved one will be forgotten.

But life pushes the living to live. "Eros" in the wider sense of its meaning, philosophically or psychologically, signifies "the life force."[5] Eros is life's urging—it's the push to continue your life. What non-grievers find hard to understand is that your life now is without your dearest one. And *you* are the one who will know when your sorrow has calmed.

Tears as a Physical and Emotional Release

However the body internalizes grief as it incorporates new routines, it also externalizes its sorrow. The survivor can become agitated or socially withdrawn, feel restless, fall into reveries of the loved one, or continue to cry.

A woman cried so long after the death of her husband that some of her relatives told her to stop, that she had cried too much and should be getting over it: "It's time to get on with your life." But when others in her family who had grieved saw her tear-streaked face, they understood that those rivulets of tears were her body's way of releasing the pain of grief. As Washington Irving wrote, "There is a sacredness in tears. They are not the mark of weakness, but of power. They speak more eloquently than ten thousand tongues. They are the messengers of over-whelming grief, of deep contrition, and of unspeakable love."[6]

Even non-human mammals, including the largest of land mammals, elephants, seem to shed tears when appearing dis-tressed and overwrought after the death of one of their own. Be-cause elephants can't tell us how they feel, what we see as grief-related behaviors, including trying to cover the dead, cannot be confirmed. Yet elephant researchers, wildlife film-makers, and others have reported how deep pachyderm grief appears to be.[7]

For humans, tears are not all the same. They vary in the chem-icals they contain, and what causes them. There are three basic kinds of tears: basal, reactive, and emotional. Basal tears are the ones that keep the eyes moist and lubricated, constantly protect-ing them from drying out. Reactive, or reflex, tears clear out the eyes from irritants such as environmental dust and vapors from cut onions. But when someone convulses with laughter, shud-ders with sadness, or is astounded by joy, and tears start to flow, these are emotional tears. And when feelings are so deeply felt that we become nonverbal, emotional tears can pour out sud-denly and uncontrollably.

Droplets that gush or stream down a face—emotional tears—

are unique in that they contain protein-based stress hormones: adrenocorticotropin, prolactin, and leucine-enkephalin. Leucine-enkephalin in emotional tears is related to endorphins, and is a natural painkiller. Part of why we feel better after crying is the effect of leucine-enkephalin.[8]

Ella Freeman Sharpe was a British psychoanalyst whose early work in literature shaped her psychoanalytic focus on sublimation, the body, and a sense of what was concrete.[9] One of Sharpe's analytic patients who had experienced the death of a loved one realized how important tears were for grieving her deep loss. The patient wrote a verse expressing her need to be left alone to cry: "Leave me my grief. Thus, undisturbed/By clamorous help, I still may weep./While tears can flow/Love is not dead."[10]

Another patient of Sharpe's, who had suffered a similar loss and was grieving, "dreamt that she was in deep water. The water however was so briny that it held her up and she knew that there was no fear of drowning." In talking more about the dream, the woman's "association to 'salt water' was immediately 'salt tears' and then the next moment the patient quoted the lines: 'Let Love clasp Grief, lest both be drowned.' "[11]

The salutary effect on the body of emotional tears—their biochemical composition, and the way that crying emotional tears reduces stress—has been a focus of study for over thirty years.[12] To cry after a loved one's death is generally healthy and beneficial.

Tears streaming down a face may glisten if light hits the droplets from a certain angle. Whether they simply flow or if they gush down cheeks, tears dry into patterns that are visually unique depending on the feelings that propel them. The shapes, patterns, and textures of tears caused by different emotions—such as sadness, laughter, and grief—were photographed by Rose-Lynne Fisher. Her technique was to drop tears onto slides and let them dry, either naturally in the open air, or under a coverslip. Each slide, with its unique pattern of a dried tear, was then placed under an optical microscope and captured as a photograph. In her book *The Topography of Tears*, Fisher published more than a hundred images of these tears that had been shed by herself and others.

Fisher clearly states that her images are "not a controlled scientific study," and lists many of the variables that could have influenced the results: "the volume of tear fluid, evaporation or flow, biological variations, microscope and camera settings, and how I process and print the photograph."[13] Even so, it seems that those tears shed in grief showed a different pattern than tears from other emotions. Several of Fisher's photographs of tears are shown here: tears of grief, tears of change, and tears of possibility and hope.[14]

Effects on the Immune System

The body maintains its health and well-being through our neural, cardiac, and immune systems. The brain controls everything we experience internally and externally. The heart pumps blood— our life-sustaining liquid—and keeps it flowing through vessels. And the immune system is the gatekeeper and bouncer, identi-

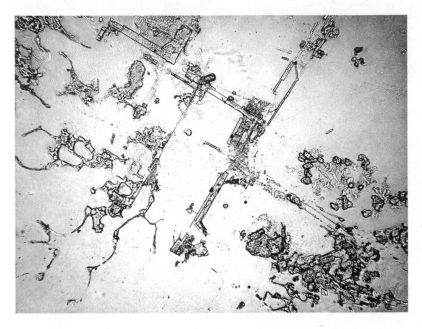

Tears of grief. From *The Topography of Tears*, © 2017 Rose-Lynn Fisher,
published by Bellevue Literary Press.

fying and destroying bacteria and viruses that could cause temporary, debilitating, or fatal illnesses.

For decades, the prevailing view in neuroscience was that the central nervous system was isolated from the immune system by the blood-brain barrier. But that has changed; research has shown that there are immune cells—T cells—in the brain that may protect against neurodegenerative and neuroinflammatory diseases.[15] The long-held tenet in neuroscience that the immune system and the brain function independently has been overturned.[16]

Previous studies, too, had found indirect communication between immune cells and the brain. Researchers learned that im-

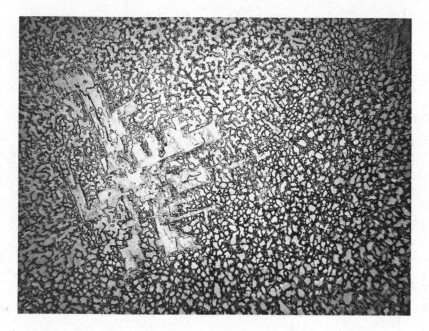

Tears of change. From *The Topography of Tears*, © 2017 Rose-Lynn Fisher, published by Bellevue Literary Press.

mune cells in the cerebral spinal fluid come from specialized cells in the choroid plexus, located in the brain's ventricles, as well as from immune cells in the meninges of the brain.[17] These early data suggest that immune cells not only influence brain functions, but also help maintain overall health in the body. In mice, studies also show that T cells are important for learning and for normal social interactions.[18]

Taken together, the results of these studies suggest that grief-related neuroimmune interactions within the body may affect the health of a survivor more than was previously thought, and may be one reason why many bereaved people experience phys-

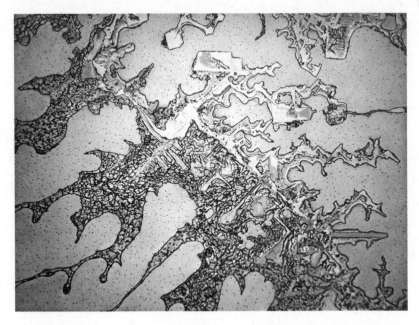

Possibility/hope. From *The Topography of Tears*, © 2017 Rose-Lynn Fisher, published by Bellevue Literary Press.

ical illness in the months following a death. The shock of grief and its related stresses (including memories of previous experiences of grief) can stress and even weaken the immune system, overwhelming—at least temporarily—the surveillance and protective functions of the brain's T cells.

If we know that the chances of becoming ill increase after the death of a loved one, what can survivors do to support their immune systems and maintain their health? Some ways to reduce the effects that grief has on the body include expressing grief and its accompanying emotions, knowing that one's appetite and sleep patterns can be affected, and realizing that one's thinking isn't the same as it used to be. Knowing what's happening to one's body, and being patient about how grieving can be draining physically, may reduce the burden on the immune system and lessen grief's stress.

Research about the effects of stress on the body will continue to expand our understanding of how grief tests the body, but meanwhile, for the bereaved, the benefits of time-tested recommendations are well worth remembering: exercise foremost, and sleep; then a healthy diet, social contact, and enriching and pleasurable activities.

A Loss after a Loss

But what about when the immune-system defenses of the body are overridden, and something happens that may seem like a physical illness, but is not? This may be a time when a survivor's psychological defenses, driven by grief, derail the body's natural ability to function protectively.

A twenty-seven-year-old mother with a previous mysterious physical problem was referred to me, ostensibly for depression. Mrs. M was soft-spoken. She looked at me in a slightly skewed way, with her head tilted. In our first meeting, she told me that her baby cried—she thought a lot, but not according to her husband. After a long pause without talking, she haltingly said, "My baby died." Tears streamed down her cheeks, which was the only hint that Mrs. M was crying because there was no sound at first and little change in her facial expression. She described what her baby had looked like: strawberry blonde hair and blue eyes.

Mrs. M said she was able to function after the funeral, which surprised her. She could make the bed, do the laundry, and perform other household tasks. But she said that after several weeks she felt something was physically changing inside her. The first time she woke up and heard no sound in the house, she remembered that Tessa had died. She thought, of course there would be no sound. There was no baby to make any sounds; there was no Tessa. It was then she thought she was beginning not to hear clearly. She said it was strange and not strange at the same time. Her hearing had changed, very slowly. She began to hear less and less clearly, and voices became muted. She found it curious that she still understood what was being said. It didn't panic her, it just seemed odd. It was as though she was hearing at some distance from the person speaking. She said the experience was surreal.

Doctors referred her for tests, looking for anything that might have compromised her hearing, such as a viral infection, but the results came back negative. During that time, she said, no one asked about her baby who died.

In one of our sessions, Mrs. M looked at me with eye-riveting intensity. She said she knew how strange it sounded that one day she realized she could hear normally again. It was like waking up from a dream. But what happened was not a dream. Not being able to hear clearly for several months was real, and the death of her baby was real. She said she knew that her hearing and Tessa's death were separate, but for her there was a connection. The awareness happened as she began to realize that even if she could hear normally, she would no longer hear Tessa because Tessa had died. It was then that she became aware of the connection between the two, and her hearing gradually returned to what it was before Tessa had died.

Mrs. M's primary-care physician had suggested that she consider psychotherapy given how much she'd been through and the mystery of how her hearing was affected. After several sessions, she decided to continue in therapy because, she said, it was a safe place to describe the severity of her pain. She said that often her grief was so wrenching, it was almost a physical pain. She kept visualizing her baby's smiles, remembering Tessa's gurgling sounds of delight. All of those experiences, she whispered, were becoming memories of Tessa's life that wouldn't go beyond babyhood.

We eventually came to the point in her therapy when Mrs. M's inquiries into her grief and hearing loss made sense to her. She came up with her own insight, one that I silently marveled at. An epiphany, she called it. She had to shut out much of the auditory world, be unable to hear, for months after Tessa's death because she had to find a way to handle not hearing her baby's laughter, her baby's crying. She didn't hear fully for as long as she couldn't accept the death of her baby. When she was able to admit her baby had died, her hearing returned. As she put it, when her hearing came back, her grieving began.

Mrs. M's grief and the pain were still there whenever she thought of Tessa. And she knew it would never completely go away. But her measured way of talking about Tessa's death gradually changed. She was able to describe details, something that in the beginning sessions she could do only briefly.

Our sessions became less frequent, and Mrs. M began to talk about going back to work. Family and friends were urging her to do so. She talked about options: whether to return to marketing or pursue something entirely different.

Sometime during one of our last sessions, Mrs. M brought up a memory that hadn't surfaced earlier. She had grown up next door to a retired couple, whom she would often stay with when her mother, a nurse, was unexpectedly called in to work for an afternoon or night shift. The husband of the couple had lost the bottom half of his left leg when he was in the military. The way he could get around using his crutches riveted her attention as a young girl. He told her it had taken him a long time after he lost his leg before he could sit, stand, and walk the way he did. He also told her that he could still feel sensations in what was the missing limb, even years later.

She had recently thought of that missing-limb sensation her neighbor had described so long ago, when she was just ten years old. She said that now, at last, as a grieving mother, she understood it. We then discussed the "phantom limb" phenomenon, which afflicts about 80 percent of those who lose a limb by war, accident, or amputation. When a limb has been amputated, some of the wiring in the brain changes so that the missing limb is felt, or experienced, as though it were still there.[19]

Mrs. M said she thought the death of Tessa was like that. Like her neighbor who continued to feel sensation in his missing limb, she continued to feel the pain of missing her baby. At first, she thought the pain would never stop. Now, more than two-and-a-half years later, it was not as implacable as it once was, but it was still there. Whenever something reminded her of Tessa, she felt her grief surge again. The dates of Tessa's birth and her death, along with anniversaries of other losses, continued to activate outpourings of sorrow.

Mrs. M began to describe sweet memories of Tessa: how it felt to hold her baby, how Tessa had bounced when she was happy, and how she had cried when she was upset. She remembered the plump softness of her arms, and how Tessa would giggle at kisses. And we talked more about grief—how it swells, how it recedes and becomes quieter, but how it can reemerge during those anniversary times or when anything else reminds her of Tessa.

There were several closing sessions. Mrs. M wanted to have a choice of which session would be the last depending on whether there was anything more she wanted to cover at the end of our work together. She told me about changes that had occurred in the house. She and her husband had donated all of Tessa's furniture, then had the baby's room painted and refurnished, turning it into a guest room. The visual impact of the room—colors, furniture, carpeting—was now strikingly different: one way, she said, that she would no longer be reminded of Tessa.

At the end of our final meeting, Mrs. M stood up, as usual, several minutes before the session was over. But she didn't pick up her things this time. Instead she walked toward me with open hands and clasped mine. She held my hands firmly as she thanked me for the work we'd accomplished together. We said goodbye, our eyes blinking from tears about to spill.

Masked Grief

What Mrs. M experienced on a physical level during the time when her hearing diminished is called masked grief.[20] After ruling out a physical cause of her diminished hearing, it became clear that what she had experienced on a psychological level was a mild form of "depersonalization," which is why she described the experience as "surreal." This bereaved mother had to detach herself from all sound because she couldn't bear *not* to hear the sounds of her baby. Not being able to hear at full capacity was the path Mrs. M's sorrow took through her body.

When the shock of a death overwhelms the survivor, somatic grief can have other effects on the body. As discussed earlier, a heart can experience "broken heart syndrome"—a transient disorder that usually resolves quickly, though in the rare instance that the survivor's heart has been weakened by previous cardiac events, can be fatal. Grief may also affect a person's sense of taste, interfering with the ability to experience flavor and so taking away the pleasure of eating. Or a grief-stricken person may experience physical symptoms similar to those of the deceased, called a "facsimile illness."[21] A woman reporting a pain in her shoulder like what her mother lived with for years may not realize the connection, though others who knew she didn't have this symptom before her mother died will recognize it. To experience similar physical symptoms, or similar ways of walking or talking, is a way to identify—albeit unconsciously—with the deceased. The similarity in physical symptoms, behaviors, or mannerisms can be a reminder of their lost loved one. Though the survivor can no longer embrace their loved one, cannot see them or hear their voice, experiencing or adopting similar physical characteristics creates a connection, extends an attachment.

It is hard to unravel physical symptoms that are part of masked grief. But there are clues. Getting sick around the time of an anniversary of the death, or a wedding anniversary, or the loved one's birthday, can suggest masked grief. Meeting someone who looks like the deceased can elicit a strong reaction. A scent—perfume or aftershave, or a long-forgotten taste or cooking aroma—can trigger symptoms of masked grief. The sense of smell is the quickest and most direct of the five senses to reach the brain: olfaction doesn't pass through the relay station, the thalamus, the way the other senses do before reaching the cortex. Proust's reveries of times past—his madeleines—are a literary testimony of this powerful and direct connection between taste, which almost always involves scent, and memory.

A man with a debilitating headache, or who is feeling sleep-deprived, after his brother's death may be experiencing these physical symptoms as a part of his grief. His sadness is real. Yet as sometimes happens with the loss of a sibling, the disenfranchised form of this man's grief may have been displaced as a physical symptom. As will be explained in more detail later, sibling survivors often feel they don't have permission to grieve the death fully because the loss seems to belong to the primary grievers such as parents or a spouse. So the experience of something physically wrong such as a headache may instead take center stage: rather than an emotional grief response, the focus is on how uncomfortable, how painful, a part of the body has become. A physical illness or a succession of ailments, one replaced by another, can become a cycle that continues for weeks, months, even years.

In a world destabilized by death, the survivor's internal and external landscapes have come apart. And grief, death's aftermath, has to find a way to be expressed, to be experienced. If not allowed its natural emotional outlet, grief can seep into the body and present itself as an ailment, a physical symptom, or a facsimile illness. Grief is unpredictable, and it will go wherever it finds an outlet. If it can't be expressed emotionally, it may find expression in the body.

PART III

Lost Loved Ones

SEVEN

Mothers

My mother is dead, and I want her back.—MEGHAN O'ROURKE

What is motherness? Mothers, however portrayed—as Byzantine Madonnas, mercurial presences in fairy tales, or emancipated professionals in contemporary fiction—embody an essence, a relationship, that is hard to capture with mere words. After birth, a baby listens, stares at, and smells the unique scent of mother while cradled in her arms. No matter the relationship later, a mother's presence, like air, can seem imperceptible . . . until you need it to breathe. From *always there*, to *there no more*—not gone for a while, but gone forever—a mother's death is unfathomable.

Just as at birth it is the umbilical cord between mother and child that is cut, at death it is the emotional cord between mother and child that is severed forever. Even if one's relationship with one's mother began at adoption, not at birth, and even if estrangement has blown stormy winds into the mother-child relationship, she is there until she dies. No matter how a mother's life ends—peacefully, tragically, unexpectedly, or heroically—her death is the final cut.

Vigils

An adult child's vigil with a dying mother can be bittersweet. Scott Simon writes a poignant essay about how he made his experience of his mother's death public by reaching out to his existing Twitter followers. During the week he spent with his dying mother, he tweeted messages, some of which were his mother's comments about dying, that reached a worldwide audience of millions. Many responded to his "140-character messages about what turned out to be the last days of her life" with comments that offered support to both him and his mother.[1]

In his essay, Simon writes of his mother's indefatigable spirit and her boisterously marvelous life. He wanted to make his mother's death known, explaining, "We want to place the face of someone we've lost in the stars."[2] It was a comfort to him and to his mother to have so many others experience through his tweets his mother's dying days. This was an unusual death watch for a son and his mother. It was a vigil that helped Simon with his anticipatory grief.

In her book *The Long Goodbye*, Meghan O'Rourke describes how tragically and angrily she lived during the long time her mother was dying. One awful experience followed on the heels of another as her mother's cancer spread. There were times that

O'Rourke was brought to her psychological knees, such as when she went shopping with her mother for new clothes, smaller ones because her mother was losing so much weight. She writes of how her urge to recapture the beautiful mother she'd known was ambushed each time: "I was stunned by the way my mother's body was being taken to pieces, how each new week brought a new failure, how surreal the disintegration of a body was."[3]

O'Rourke's pain was amplified immeasurably after her mother died; the death became a calamity, from the Latin word "calamitas," to shout, to cry out. Nothing that she thought would help did. She was awash in her grief. Time, structure, friends—all came to a halt. The world that existed before her mother died didn't work any longer. Her mother was gone.

Young Grief

Children's grief is intense, and lacks understanding. A young child who doesn't understand death won't know why his mother doesn't return; he might even think that his mother is nearby, that she has just gone away temporarily and will be back. Grief can be experienced as physical: tummy aches or headaches; as behavioral: acting out at school; or as despair: sinking into a silent lethargy and surges of crying. Thoughts and fantasies may be clung to for a little while, but when inevitably they fail, a child can be inconsolable.

Roger Rosenblatt's heart-rending memoir, *Making Toast*, tells how he and his wife moved in to care for their daughter Amy's young family after Amy died suddenly from a rare heart condition. The author and his wife left their home on Long Island to live with and support Amy's husband and the three children.

Rosenblatt's youngest grandchild, James, whom they called Bubbies, was just beginning to talk when his mother died. At the post office one day, Rosenblatt received a call from his wife that propelled him into another level of shock. His wife told him that Bubbies had cried out "'Mommy,' as if calling her," and asked, "When is Mommy coming home?" Rosenblatt wonders whether, all this time, his little grandson has "been thinking she was simply away?"[4]

While the child seemed fine the next morning, Rosenblatt was not. He was the inconsolable one. Earlier in the book, he quotes a friend who tells him, "Sometimes there are no words."[5]

Unable to Speak

As explained in Chapter 3, early in life every child is kinesthetic, using her body to learn about the world. During this time the child has little access to words; she is too young for language, which later will describe and enrich her first impressions of the world. Establishing a secure attachment to the mother is critical during this stage of development.

The failure of words in a grieving adult can signal a retreat to this time before language, when everything that happened was experienced through the body. For many, it was also the time when Mother was *always there*.

News of a mother's death, no matter how that news is delivered, is felt in the body. From the brain, the amygdala sends its distress message to the hypothalamus, which initiates a cascade of brain responses. Thinking shuts down: executive function (the task of the prefrontal cortex) goes on hold, and the brain's automatic processing takes over. Thinking gives way to feeling. A mother's death can be followed by a gnawing, relentless searching and aching.

Left without a mother, experiencing the new state of "motherless," a survivor can be bereft, which as discussed earlier means to seize, to steal, to take away. A mother has been taken away by death, stolen by death. Yet even months later, those left behind may still believe that she'll come back. O'Rourke, in *The Long Goodbye*, recounts that four months after her mother died, there were times when she "still privately believed" her mother was "coming back." With one's emotional life thrumming below the logic of consciousness, and with unconscious links to the long-ago preverbal stage fully activated, imaginings like this are common and seem real. In that fantasy realm between death and the emerging reality of life without one's mother, a bird or a tree can become a symbol of a mother's return. Sometime after her mother's death, a cherry tree in bloom moves O'Rourke to wonder, "Why would it be here unless it was to announce the universe's mistake? *Here you go. She's a good woman. You can have her a bit longer.*"[6]

A Daughter's Ambiguous Grief

On the anniversary of the day she heard the news, my grandmother would often tell me what happened to my great-grandmother. I'm not sure how much of the story is true: the only photo my grandmother had of her mother was a black-and-white print, frayed at the edges like my grandmother's memories. The identifying feature of my great-grandmother was her unusual height: during the late 1880s, she towered over everyone in the village. My grandmother said you could see her head from anywhere in the church; she was almost a full head taller than everyone else. In that faded, grainy photograph of her and my great-grandfather, her kerchiefed head is higher than his, and a smile edges her lips. His expression is taut and somber.

In 1900 my grandmother, a young woman, arrived in the United States. Nearly forty years later, a brother who had emigrated to England was her only source of any news of their mother. Soon after the Nazis invaded Poland in 1939, their seventy-nine-year-old mother, my great-grandmother, had been taken to a work camp.

After the war, with much of London in ruins and my grandmother's brother very ill, there was still no news from their village. My grandmother clung to the hope that as a woman who towered over everyone in the village, her mother's strength had helped her. But my great-grandmother was never heard from again.

Every so often, when something triggered thoughts of her mother, my grandmother would become quiet, which was unusual for her, and keep especially busy. Eventually she made up her own story of her mother's death. Occasionally she would say, "My beloved mother, my poor dear mother, died in the work camp, unburied, her body prey to insects and animals."

My grandmother was denied what is so important to grief: for the rituals involving the body of the deceased to take place—the wake or memorial, funeral, and burial. No one knew what happened to her mother, so my grandmother's ambiguous grief continued unabated.

Purposeful Cuts

Years before the death of my own mother, I was part of a tragedy involving the death of someone else's mother. Living in a first-floor apartment in a small apartment building, I became good friends with my neighbor across the hall. Together, we began to develop a relationship with a woman who had recently moved in upstairs.

The three of us met for coffee several times, and the new neighbor talked a little about her life. She had two daughters: a nineteen-year-old college student living with her for the summer, and an older, married daughter, who lived close by.

A few days after Mother's Day, my new neighbor and I chatted briefly as we walked to our cars outside. Alternately smiling and grimacing, she told me she'd gone through a difficult divorce during the past year. We reached her car and she quickly got into it, said she'd cope no matter what, and hastily drove away.

The following Saturday morning, her younger daughter, sleepy and dressed in pajamas, knocked on my door and asked to use my bathroom. I said of course, but asked the girl what was wrong with hers. She said she'd slept late and was feeling a little "spacey," and it seemed the bathroom door was locked. After she came out of my bathroom, I told her I could unlock her bathroom door; the doors inside the apartments were pushbutton-locked and could easily be opened with a small screwdriver.

My friend from across the hall, who was with me, looked at me and asked, "Are you sure it's a good idea to try to unlock that door?"

I thought she meant did I really think I could unlock the door? I said, "Oh, it's okay, I know how to unlock these doors."

"I don't think that's a good idea," she said, but I was already on my way up the stairs, screwdriver in hand.

The apartment door was open. All the blinds were drawn and the apartment was dark, in stark contrast to my kitchen downstairs, which was filled with the bright morning sunlight. Nearing the bathroom, I smelled something that made me think perhaps the woman had fainted on the toilet.

It was easy to open the bathroom door with the small screwdriver. The light from the bathroom window lit up the small room. In the pink tub just like mine downstairs, I saw my upstairs neighbor. She lay in the bathtub in a short-sleeved bathrobe, her left foot sticking out of the tub, white and stiff. The cut on her left arm was vertical, from her wrist to her elbow. The room was hot, the air thick and heavy with the slightly sweet smell of her death, hours old.

I did not see her face. In shock, I couldn't do anything but drop the screwdriver and run out of the apartment.

Her daughter was downstairs in the hallway with my friend, asking why the door was locked. Trying to keep my voice steady,

I told her that her mother had had an accident, and we needed to call her older sister. The girl gave my friend the number, and my friend gently guided her into my living room. In the hallway, I told my friend what I had seen and advised her to call the police and the girl's sister.

I was beginning to feel faint. I tried to keep moving and my friend told me to go into her apartment across the hall and lie down, that she would take care of the daughter and make the phone calls. Later she told me that I began to cry, then immediately fell asleep. The next thing I remember is waking up when the police arrived.

The police questioned me briefly, then went up to the apartment. When they returned, one of the officers, who seemed dazed by what he'd seen, asked where the woman's purse might be. I suggested the bedroom. I didn't see the medical team who came to take the woman's body away. The older daughter took her younger sister home with her.

Later that week, the older daughter came to speak with me in my apartment. She told me their mother had lost her job the previous week, and seemed depressed. But to do this, and while her sister was asleep in the apartment? It was hard to believe, impossible that this could have happened. She was dumbfounded. She said her younger sister would be coming to live with her and her husband until her sister went back to college. Both she and her sister would be going into counseling.

My reaction to what I saw that day stayed submerged for years. The trauma of discovering the suicide of a neighbor I barely knew, a mother, was inaccessible until after my own mother bled to death some years later.

My Mother's Ides of March

Even when our grief about a long-ago death seems to have faded, something can happen that will make it reemerge unexpectedly. Grief can surface in dreams, be displaced into physical symptoms, and cripple relationships, because it never goes away completely, it just stays dormant. Julian Barnes writes of grief: "We may think we have beaten it, when it has only gone away to regroup."[7]

March 15 was my mother's birthday. So many times when I was growing up, I would hear my mother say: "I was born on the Ides of March, the day Caesar was killed." For me, the day of her birth became a haunting link to the death of Caesar. Just as Caesar's death was foretold, so it seemed that my mother predicted her own death.

In Shakespeare's *Julius Caesar*, Caesar hears the prediction, warning him three times:

CAESAR: Who is it in the press that calls on me?
    I hear a tongue shriller than all the music
    Cry "Caesar." Speak. Caesar is turned to hear.
SOOTHSAYER: Beware the ides of March.
CAESAR: What man is that?
BRUTUS: A soothsayer bids you beware the ides of March.
CAESAR: Set him before me. Let me see his face.

"On the Ides of March" denarius, depicting
Marcus Junius Brutus on one side, and, on the other, a pair of daggers
and liberty cap representing the assassination of Julius Caesar.
Classical Numismatic Group, Inc., http://www.cngcoins.com.

CASSIUS: Fellow, come from the throng. Look upon Caesar.

CAESAR: What say'st thou to me now? Speak once again.

SOOTHSAYER: Beware the ides of March.

CAESAR: He is a dreamer: Let us leave him. Pass.[8]

On that fateful Ides of March day, Caesar is stabbed by his senators and bleeds to death on the steps of the Senate. How was my mother able to connect the date of her birth, and the date and method of Caesar's death, to the way she herself would die? Like Caesar, she bled to death. She endured a procedure meant to restore circulation in her leg, but it went very wrong. Caesar died at the hands of his assassins, but my mother died undergoing a medical procedure intended to help: although sedated, she somehow managed to pull a stent out of her arm not once, but twice.

My husband and I visited my mother the night before her procedure. We also talked with her cardiologist, who was confident and optimistic. A telephone call to me the next afternoon was the first hint of trouble. My mother had pulled the stent tubing out of her arm and could lose her hand if the problem was not corrected. "Give the team an hour," the cardiac fellow suggested. We didn't wait, but headed to the hospital. Calling before we left, the news was that things were "back on track."

At the hospital, the doors to the special procedures room were open. Even though we were told not to go in, I caught a glimpse of my mother on the table. The attending physician said all was proceeding on course, and they would call us in several hours after the procedure was over.

We drove to a cousin's who lived very close by. As we walked in, she told us the hospital had called to say that the situation had turned critical.

We drove back and went immediately to special procedures. No one stopped me this time. The double doors opened, and I walked in to see blood on the team's shoes and on their scrubs. Although my mother's arms had been strapped to boards, she had managed to pull out the stent again. The chalky color of her face was in stark contrast to the blood on her arms and the wooden supports to which her arms were fastened. Walking closer, I heard her nearly inaudible moans.

I bent down to her and said, "I'm here."

Her response, my dying mother's last words, was "Let them let me die."

And I did. I told the team to stop. They wrapped her and wheeled her out of special procedures. My husband and I were told to wait until they had her settled back in her hospital room.

In a waiting area, I paced back and forth—innumerable times, it seemed. To my husband, who was sitting down, I said, "She's dying."

When we went into her room, it was then that time stopped for me. There was no time. As it had been so long ago with my mother and me, she and I were in a space where time, and even words, did not exist. I was there again with her where there was no conscious thought. It was just me and my beloved mother. Mommy and me together, alone, and I remembered her smiling, how at times she would beam at me, proud of me and delighted to be with me. And the bond from that long-ago time when all that mattered and all that existed for me was my mother: that bond was breaking, that time was ending. I stroked her face, crying, my heart breaking as her heart beat slower and slower. Then her heart stopped beating. The monitors stopped. She was dead. My mother was dead.

It was as if something had fallen off the Earth. My mother's life was over. I stayed alone with her for a time, stroking her forehead, and crying tears that had no name. I left with a lock of my mother's hair, cut with scissors a nurse had given me.

Unexpected Cuts

Some years later a patient began therapy with me who also had experienced a trauma related to her mother's death. Diana was in the third month of her sabbatical from teaching and at our first meeting, she cried, telling me that her mother had died three years earlier, and that everything associated with her mother's death had been, and still felt, horrible. She described how what had happened was unusual and wondered aloud what about death was usual. She had a way of talking around the topic of her mother's death.

Her mother had died unexpectedly while Diana and her husband were on vacation in Hawaii. Due to a communication glitch, she didn't learn about the death until the next day, and another twenty-four hours passed before she was able to return to the mainland and continue on to Maine to identify her mother's body. From that time on, whenever she thought of her mother, she could not stop seeing those final images of her mother's body in a black bag.

At some point after our first meeting, I asked Diana if she could try to describe what happened that had made her mother's death so horrible. I gently urged her to put what happened into a narrative. She shook her head, telling me it was too painful to talk about.

These were the same words she would use later to describe how both her therapy and her sabbatical were going. The two activities seemed at times to run parallel to each other, and it was remarkable that she was able to commit to both with such intensity.

Diana and her mother had been close. Perhaps too close, she thought. She said that they hadn't grown apart as she grew up. Further, when looking back at the relationship, she realized that when she had married and moved out of state, she had compartmentalized the pain caused by the separation from her mother.

Now, however, she was truly motherless, and she felt both angry and horrified.

When she was able to describe what happened, Diana said her mother had died in a rather bizarre accident. Friends in the neighborhood told Diana that one day, at dusk, her mother was out walking when a man on a bicycle, on the sidewalk, was riding toward her. Neither apparently saw the other. The bicycle crashed into her mother, knocking her into the street and killing her. The body was held by the medical examiner until Diana, her mother's only close living relative, could identify it.

Diana then described reaching her mother's body a full two days after she had died. There were so many parts to the horror, she said. She described the experience of identifying her mother's body as gruesome, and said she would never forget the sound of the body bag being unzipped.

The effect on Diana of her mother's death had put a strain on her marriage. She and her husband had been working things out, and it was at his urging that she had decided to try therapy.

The trauma she had endured while identifying her mother's body had prevented Diana from accessing her grief. She hadn't been able to distance herself from her experience in the morgue. But in our sessions that began three years later, the words that she used to name her feelings and tell her story made real what had been so awful for her. She said she felt as though she'd awakened from a long nightmare, and was finally grieving her mother's death. She described how much she'd loved her mother, and talked about how much she missed her. Her mother's death was no longer saturated, as it had been, with gruesome memories.

Diana finished her sabbatical successfully, completing the book she had begun during her time off. Therapy had helped, she said. She had learned how powerful words could be—in writing her book, and in naming her feelings and talking about them. Her words had given her the ability to convert the trauma of her experience into grief.

Natural Cuts

Another patient came to see me. Marnie was thirty-five, unmarried, and a successful lawyer. She had decided to begin therapy because she wanted to understand why she wasn't enjoying life.

Marnie seemed surprisingly at ease during the consultation. Medium height, slim, well-dressed, with thick black hair, she explained that her only sibling, a younger brother, taught math at a private secondary school and both her parents had died. Immediately after finishing the initial paperwork, she asked in a challenging voice if I thought therapy would help. Then, rather testily, she added that the previous therapy hadn't helped after her first committed relationship ended.

Quickly recovering her poise, she said that her family history was the eternal maternal drama. It was the same play, same actors, only performed at different times in different theaters. Her mother had been an alcoholic, yet amazingly could function beautifully—outside the immediate family. Marnie became the fixer in the family, a role she told me that as an adult she was paid to perform in her profession as a lawyer. She chuckled a little.

As therapy progressed, Marnie seemed to move easily back and forth between the present and the past. At one point, she remembered that she'd hidden her mother's car keys after she saw her mother add something colorless to her own orange juice, and wouldn't let Marnie taste it. It was raining that day, she said, but she told her mother that she wanted to walk to school. Her mother hadn't even noticed how hard the rain was falling.

In a later session, she said that growing up she became the fixer-in-training and missed the fun and spontaneity of childhood. It made her distrust just about everyone; whatever anyone told her, she had to find out for herself. Always having to monitor her mother's mood made her hyperalert, which interfered with getting close to other people. But at school she found refuge and success. Academics became her lifeline. She excelled in school, and now in her profession. She depended only on herself.

As a child Marnie was ashamed of her mother, not knowing when her mother might begin a downward spiral. She never invited friends to the house; she didn't want her schoolmates to know the secrets of her family: her mother's drinking, and her emotionally unstable relatives who would sometimes stay over. For the relatives, her mother was the fixer in the family, and Marnie had inherited the same role. (A generational repetition, I suggested.) She remembered uncles and aunts who would stay for several days. They depended on her mother to help them.

She was their heroine. But not Marnie's, since after things for the family were put right by her mother and the relatives had left, her mother's drinking would begin.

During one of our sessions, Marnie described not understanding why she was feeling terrible, more than just a general malaise. I, too, wondered what was going on and asked her to tell me more about the deaths of her parents. She said she'd already told me both her parents had died, her mother just two years earlier. Then she remembered that it was the week that her father had died, of Parkinson's, ten years earlier.

We discussed that an anniversary of a death, without conscious awareness, can stir up grief in unpredictable ways. As volatile as her mother was, her father had been a calm and dependable presence in her life, and she missed him terribly. Marnie's smile was lovely, and her voice was soft, as she spoke of him. Talking of both parents, her voice alternated between sadness and anger. She would sigh, shake her head, and look down.

Marnie described how her father had become weaker and weaker as the Parkinson's had progressed. She explained that his gradual decline made her feel powerless and angry. Like her mother she'd been the fixer, but there was nothing she could do to fix her father. She talked of how much she admired the way he dealt with life's ups and downs. With his Parkinson's disease, he wanted to know the details of his prognosis. She said that being able to predict what would happen helped him to cope. He had been a conductor for the railroad, accustomed to schedules and routines, and he liked knowing what came next.

Marnie told me her mother would snap at her father when he'd spill his tea because he couldn't hold the cup steady—and that it was hard for Marnie not to lose her temper in this sort of situation. Her mother's behavior toward her father wasn't new, for she had regularly criticized him even before he became ill. But it made Marnie wonder how her father put up with her mother's criticism of him, how it didn't seem to affect him. At one point during his illness, she said, her father pleaded with her to be kind to her mother because he understood how hard it was for her mother to see what was happening to him.

Two years after her father died, her mother had a stroke. She had been living alone, and needed more care than Marnie and her brother could manage. When she and her brother suggested that an assisted-living home would be better for her, her mother cried uncontrollably for days: she did not want to give up her home and go to live in a place where everyone was old.

But move she did. Marnie and her brother took turns visiting her, and to Marnie's surprise, her mother calmed down after she moved. Spending time with her mother turned into a delight for Marnie, who described how she looked forward to their visits. She found a new side to her mother that had not been evident earlier in their relationship, and she spent more and more time with her. As she described to me how the relationship changed, her voice took on a softer, gentler tone. They had learned to enjoy each other: laughing at times, and remembering times when her father would humor her mother, but lovingly.

Then, unexpectedly, her mother had another stroke and died.

Marnie couldn't believe it; she said she didn't want to believe it. She felt devastated at her mother's funeral. Her brother remarked that he thought she seemed more upset at their mother's death than their father's. She told him she felt cheated. How could this new relationship with her mother, these warm and intimate times together, be cut short?

As Marnie grieved and cried, we talked about grief, and how it manifests differently with each loss. She said that though she had grieved for her father, after her mother died she had felt a weight in her heart that wouldn't let her enjoy anything. Now her mother was gone, and she was an orphan. She felt abandoned and alone. She found it impossible to talk to her brother, who was busy with teaching and his family.

Sometimes during a session, she sat quietly, dabbing at tears. She had begun to cry, but only during therapy, she said, where she felt it was safe to cry. After several months, she began to inhale and exhale with effort when she would cry and talk about her mother. She wasn't able to find the words to describe her emotions.

Sometimes, she would look at me beseechingly and ask what words she could use to describe what she was feeling. At other times, she would glare at me, and ask why the words were so hard to find. I explained that the stronger the feeling, the harder it is to find words, and that words to describe grief and other intense emotions can be elusive in the beginning. But to find the words that capture intense emotions is a way to manage feeling overwhelmed, and putting feelings into words, I told her, can help manage those feelings by changing the way the brain processes them.

Marnie listened, and talked more, and gradually she described feeling angry, guilty, and profoundly sad. And she talked about how much she missed her mother.

She cried a little less.

She continued to express regret that she and her mother were unable to enjoy each other. It made her angry that her mother had died so suddenly. And she felt guilty that she'd been so hard on her mother for so long. She wondered if her mother-grief was permanently linked with mother-guilt. Was the loss of a mother ever separate from guilt? Are these emotions inextricably linked with a mother's death?

She started to remember times that were happy, during her early childhood, before her brother was born. She didn't think her mother drank then. When they'd go shopping together, she'd buy Marnie pretty clothes, and books, and they'd go to lunch.

Marnie's grief after her mother died had been delayed, stalled by many other feelings such as anger, guilt, and remorse. Although it took time to articulate the complex relationship she'd had with her mother, finding those words allowed her grief to be experienced: it flowed, surged, receded, pushed back into consciousness, and then quieted. As her sadness and sorrow became easier to identify and acknowledge, Marnie began to move on with life.

To lose one's mother is to lose the self that once was with her. The lifeline that began with the umbilical cord—or with an attachment surrounding adoption—and developed into a deep emotional bond, has been cut. Yet a mother's quintessence, the essence of her, internalized first as the child's idealized image and later as the adult's more realistic one, can be a source of strength, even in grief. A mother's very motherness can help mend her son or daughter's breaking heart, even when she can't be there anymore.

EIGHT

Fathers

The fabric of the world has torn.—HELEN MACDONALD

What effect does a father's death have on a son or a daughter? Fathers die in war, as soldiers, marines, airmen, sailors; others die close to home in the line of duty, as police officers and fire-fighters—those fathers who protect people outside their family often become heroes in the community and the nation as well. There are fathers who are heroes inside the family, too; private heroes to their children who feel their loss deeply: daughters and sons left with lionized memories wrapped around the black hole of grief that can follow their father's death.

And then there are the literary fathers: King Lear wreaked his advancing madness on a daughter who defied him, while Hamlet's dead father commanded him to avenge his murder. In

Michelangelo's Sistine Chapel, God creates a magnificent Adam simply by extending his finger to breathe life from nothing into something. The power and majesty of the father occupies a place in myth, religion, art, and literature.

Oedipus

Long before Sigmund Freud conceptualized the "Oedipus complex," the myth of Oedipus existed in Greek legend. Sophocles wrote the play *Oedipus Rex* in the fifth century BCE about the tragic king Oedipus.[1] When Oedipus sets out to end a plague on his city of Thebes, he learns that unwittingly he has killed his father and married his mother, as had been predicted by the Delphic Oracle.

Over two thousand years later, Freud, steeped in the intellectual milieu of Vienna and under the constraints of Victorian mores, took the tragedy of Oedipus and refashioned it to fit his theory of the mind. In 1897, a year after the death of his father, Freud told his colleague, Wilhelm Fliess, that "the oedipal relationship of the child to its parents was 'a general event in early childhood,'" and went on to assert with certainty that "it was an 'idea of general value' that might explain 'the gripping power of *Oedipus Rex.*'"[2]

Four years later, in 1901, Freud would write about the meaning of a father's death in *The Interpretation of Dreams*. As Adam Phillips interprets these passages, when Freud writes that "the most important event, the most poignant loss, of a man's life" was his father's death, Freud is also expressing how profound was the effect of his own father's death.[3] When Freud's father died, Freud wrote to Fliess: "By the time he [my father] died, his life had long been over, but in my inner self the whole past has been reawakened by this event. I now feel quite uprooted."[4] During those years, 1897 to 1899, Freud was intensely involved in self-analysis. The death of his father and the myth of Oedipus exerted tremendous power over his thinking, greatly increasing his sense of loss. The myth of Oedipus became embedded as deeply in the ethos of Freud's psychoanalytic values as love and work were embedded, and indeed, was at the center of Freud's work.[5]

Freud laid out his theory of the conscious and unconscious workings of the mind in *The Interpretation of Dreams*, calling it a topographic system.[6] In the topographic system of the mind, Freud theorized in 1900, the conscious part of the mind was accessible and understandable. Yet underneath the rational and the conscious bubbled the unconscious, a hidden, inaccessible part of the mind, as incomprehensible as dreams. This became one of two enduring systems of the mind that would penetrate Western culture: the second, which he published in 1923, was the structural system, which included the ego, id, and superego.

According to Freud, it was from the unconscious that self-defeating behaviors and puzzling physical symptoms arose. It was only through these disguised ways of functioning that unacceptable feelings, which were being repressed, could be expressed. That is, no way could anger, competition, and jealousy toward a father or mother be expressed directly; this cauldron of unacceptable thoughts and feelings would remain locked in the unconscious until accessed by psychoanalysis. It was through Freud's method, called the "talking cure" by one of his patients, that forbidden topics and feelings—like rage, or rivalry—could be brought into consciousness by being described and discussed in a nonjudgmental environment. By naming and feeling what was hidden, that which was defended against was brought up to consciousness, exposed, and defused.

Many years later, neuroimaging studies showed how naming negative feelings can change activity in two areas of the brain, making emotions less intense. Using fMRI, Matthew D. Lieberman and his colleagues showed that putting feelings into words, called "affect labeling," led to a decrease in activity in the amygdala, and an increase in activity in part of the prefrontal cortex.[7] When participants in the study viewed photos of faces that looked angry or fearful, activity in the amygdala increased, but when the participants used words to name the observed emotions, activity in the amygdala decreased and activity in the right ventrolateral prefrontal cortex increased. In other words, the subcortical region that responds to potential danger and is associated with negative *feelings* was calmed, while the cortical region that is associated with *thinking* about emotions was engaged, all because of what makes us human—language.[8]

But suppose a child is discouraged from expressing certain feelings. If a boy has been taught not to express "negative" emotions, such as anger and jealousy toward his father, the feelings will become submerged below the level of consciousness. In these situations, something like competition with the father may be experienced later by the adult as a lack of ambition. And the mandate not to talk about negative feelings becomes second nature: situations that threaten to disrupt the established equilibrium of the family are avoided.

The developmental trajectory of Oedipal issues can be problematic for children of both sexes. For a man whose competitive and angry feelings toward his father have been quashed growing up, and who therefore has had a troubled relationship with his father, the father's death marks the time when direct competition, or lack thereof, is no longer fueled: gone is the impulse to argue (the fight instinct) or to withdraw (the flight instinct). Yet those competitive feelings can reemerge in oedipally charged interactions with others. A daughter's situation might be similar, should she find herself vying with her mother for her father's affection, or standing up to her father in arguments with him. If her father demeaned her or verbally abused her, or otherwise made her feel insignificant, her father's death might be liberating to a degree. Yet after his death, the lingering desire to earn his admiration can turn into a search for approval in someone outside the family, someone not suitable.

When the relationship is positive, the death of a woman's father is a time of tremendous loss. The daughter will no longer experience the presence of the first male figure in her life who was encouraging and loving, and who made her feel able to accomplish anything. For a son whose father was available and inspiring, the loss of that paternal love and support can leave him rudderless.

Oedipus Redux

Freud's discourse on the Oedipus complex in *The Interpretation of Dreams*, even more than the myth from Sophocles' play *Oedipus the King*, influenced Western literature, psychoanalytic theory, and clinical treatment. Despite contemporary arguments from critics, clinicians, and others challenging Freud's theoretical premises and clinical objectives, the idea of an Oedipus complex, controversial as it is, continues to inform an understanding of the human psyche and human behavior. Even when stripped of what it can be criticized for (depending on one's perspective) in today's zeitgeist—clinical short-sightedness, misogyny, theoretical obtuseness, and more—the rivalry between father and son, with or without a theory, is ancient and real. From the time centuries ago when the Oedipal conflict was dramatized as a full-scale, five-act tragedy, to the present, this parent-child drama continues its run. For some, the conflict resolves early and well, while for others, the competitive conflict simmers below the level of conscious awareness with deleterious effects—for both sons and daughters.[9]

The main tenet of the Oedipus complex, and the reason the idea has survived from Sophocles to beyond Freud, is that the intense rivalry between father and son for the mother's affection, like any other type of competition between them, is not unusual. For a daughter, the conflict with her mother is for the father's attention and affection. This developmental stage for girls and boys begins at around age three. If the Oedipal rivalry between a son and his father, and a daughter and her mother, continues without some psychological resolution, it can develop into a perennially painful conflict.

When a father dies, a son is left without his primary rival either to argue with or to avoid conflict with. Others in the outside world can become stand-ins for a son's deceased father, either to be avoided or to be targeted for anger, while a daughter may invest her love in an unsuitable choice as well as have conflicts with male authority figures. Oedipal issues that continue to be active after a father's death will influence the grief of the surviving child, affecting her or his ability to face the pain of this permanent separation.

Some two decades after his own father died, Freud wrote his condolences to Ernest Jones on the death of Jones's father. Freud communicated his sympathy with a gentle warning about what the future holds for a son upon his father's death: "You will soon find out what it means to you," he wrote. "I was about your age when my father died (43) and it revolutioned my soul."[10]

A Vanquished Oedipus

In a clinical practice, sometimes it's clear that an Oedipal conflict between a father and a son has continued not only into adulthood, but even after a father's death.

Leonard was feeling depressed and had come to therapy reluctantly and angrily. He was tall and well dressed, his jet-black hair swept back. He spoke softly but deliberately and clearly, and the way he would pronounce some words suggested that he might have studied "voiced consonants" in an acting class.[11] He told me that because he was older (fifty-five) and had recently experienced cardiac issues, he had decided to sell the business he had inherited from his father—and to try therapy.

He had been married, but only for a year when the marriage ended. He said he really didn't know why he hadn't remarried; he would have liked to have had his own children. His two sisters were both married, and he was close to them and their families. He described a childhood filled with team sports and individual pursuits, which on the surface sounded exciting and interesting. As an adult, he was often involved in high-risk adventures. He thought it was important to keep challenging himself. This is what he and his father had quarreled about when he was urged by his father to stop participating in risky activities after he broke his femur in two places bungee-jumping. What his mother called courageous, his father called reckless. Leonard's mother had always been his cheerleader.

Yet when Leonard described his childhood, details were scarce. When he spoke of his father, his descriptions of him were flat, lacking human depth. In contrast to his sports bravado, Leonard said that he hadn't had much success romantically after his divorce. He was more comfortable being pursued. He said he hadn't taken advantage of his one opportunity to marry a second time. He and a friend of one of his sisters became engaged. It was a surprise to both of them that they fell in love and decided to get married.

He was in his late thirties when he told his parents of his engagement to Alicia. He was working in his father's insurance business, and he and his father were getting along. He shook his head and sighed heavily. He murmured it was the best time of his life: exciting and rewarding. But short-lived. A phone call from his older sister on a Saturday afternoon registered shock. He heard his sister tell him that their father had died in an automobile accident early that morning.

Much as Freud's father's death "revolutioned" Freud's life, so did Leonard's father's death radically change Leonard's life. As Leonard was growing up, arguments between his father and him would typically spiral into a larger conflict. Yet soon after Leonard joined the family business and become engaged, the tension between the two men diminished. To Leonard's surprise, he and his father had begun to develop a new relationship. Looking back, it was as though joining the business and becoming engaged again were expectations that his father had had for him.

His father's sudden death changed everything. The reality that his father was not there any longer just didn't make sense. His father hadn't been ill. Soon, life became dreary. Leonard couldn't commit to a wedding date. He became withdrawn and occasionally brusque, he said. To his relief, he and Alicia broke off their engagement. He continued half-heartedly in the business, but it didn't thrive the way it had when his father was alive. Finances began to slide, though he managed to keep the business afloat.

A few years later, Leonard began experiencing chest pains that were diagnosed as angina, and coronary angioplasty was recommended. He resisted having the procedure at first. But because his symptoms were becoming more frequent and worrisome, Leonard agreed to the procedure.

Early in therapy, he described his surprise at having a heart problem: he exercised regularly and watched his diet. He connected the blockage in his arteries to his grief, which had been blocked after his father's death. Not being able to grieve seemed to mirror his inability to change the lackluster life he was living.

As therapy progressed, Leonard was struck by how angry he felt toward his father. The feelings that surfaced were upsetting, but therapy gave him the emotional space he needed to express his grief as well as his anger. As he talked about the upsetting verbal bouts with his father, Leonard described how joking that turned into subtle insults could quickly change into angry disagreements. Over time, as he talked about his father, the words he used became less charged.

Gradually, the pace of Leonard's life picked up. He sold his father's insurance business and started his own consulting firm. Financially, he got back on his feet. He bought a house in the country and renovated it, making it a home he loved. He was pleased to realize that both his parents, not just his father, would have approved of it. He became interested in genealogy and began research to find his ancestors. These activities helped foster a newfound equilibrium that reflected his psychological growth.

For years, Leonard hadn't been able to grieve his father's death. Anger that his father had died had intensified his earlier anger toward his father, and the accumulated emotions—anger, along with guilt and remorse for the times that he and his father had been verbally combative—had inhibited a full expression of grief. It was Leonard's heart condition that allowed an opening into the delayed grief of a son whose father had died unexpectedly. This opening became the catalyst for an expression of delayed grief, a grief connected to the Oedipal-fueled conflicts that Leonard was able to rework and reexperience. During therapy, as conflicts and memories were revisited, often with strong emotion, the admiration and love that Leonard had for his father became central.

Paterfamilias

Traditionally, fathers have represented authority, masculinity, and knowledge. Fathers fixed things, knew things, and earned the money—or inherited it. Fathers embodied strength and understanding. Dad was up to the task and could always make the world right again.

In today's world, this picture of the father as the sole breadwinner and fixer feels faded and out of step. It's more common now to have a two-income parent unit and for fathers to share household chores and childcare duties with their partners, who sometimes are the family's primary earner. Given the fluidity of contemporary households, children today are growing up in all kinds of family arrangements.[12]

Furthermore, in this changing landscape of modern family forms, fathers are scarcer. Luigi Zoja notes that not only is the symbolic figure of the father fading, but statistics support the impression of the father's disappearance from the family. Given the data about this trend, Zoja states that "half of all children grow up without a father."[13] Children who are abandoned or who suffer from other sources of privation may spend a lifetime creating an idealized image of their fathers. Who he was is built on the fantasy of what they would have liked the father to be.

But even in families where there are fathers, there are some who don't measure up, who drink and rage and abuse. There are also fathers who don't measure up because their world collapsed and crushed them. How many kinds of fathers do bereaved adult children mourn? How often is grief for a lost father ambiguous and unending?

Yet there are fathers who live with the family and are honorable, reliable, and loving: these fathers Zoja calls "good-enough." And there are many of them.[14]

Sudden Death

For a daughter who has a loving and deeply meaningful relationship with her father, and whose maternal Oedipal relationship is relatively unburdened, her father's unexpected death can be a hand grenade that explodes her world. In her memoir *H Is for Hawk*, Helen Macdonald tells in glittering prose how her father's death was just that, an unanticipated tragedy that blew her life apart. She becomes entrenched in a black sorrow that lasts for months. Captivated by falconry—"the hunting of wild quarry . . . by means of a trained bird of prey"—since she was a child, she feels compelled to buy a raptor to train.[15] The lure of falconry is medieval and masculine. The bird is a wild spirit to be controlled, yet is free to soar when released.

Helen buys a young, female goshawk, and names her Mabel. As she drives home with the bird in a box, newly bought from a man on a quay, she describes how her vision of it clears. "Slowly it resolved into specks of feather-dust, little pieces of the crumbled keratin that protects growing feathers, loosed from the hawk's young plumage and lit by a shaft of stray sunlight from a crack in the top of the box. Eyes and brain fell into place, and now I could see a dull shine of half-light on one lemon-yellow, taloned foot. Dim feathers, shivering with apprehension. The hawk knew she was being watched. I shivered too."[16]

As Helen begins to care for and train her goshawk, her grief becomes immersed in a year-long life with the wildest of birds. Helen trains Mabel, feeds her and watches her weight, and finally flies her. All the while she is reading and comparing her ventures with Mabel to the experience of T. H. White, who writes of his bird in his book *The Goshawk*. There are times when, reading White's botched and painful experiences with his hawk, named Gos, she imagines White's bird lost in the dark forest. Macdonald wants "to slip across the borders of this world into that wood and bring back the hawk White lost. Some part of me that was very small and old had known this, some part of me that didn't work according to the everyday rules of the world but with the logic of myths and dreams. And that part of me had hoped, too, that somewhere in that other world was my father. . . . He was still out there, still, somewhere out there in that tangled wood with all the rest of the lost and dead. I know now what those dreams in spring had meant, the ones of a hawk slipping through a rent in the air into another world. I'd wanted to fly with the hawk to find my father; find him and bring him home."[17] Her grief is taloned, vicious, and impossible. Her longing for her father is an unbearable burden that is offset—sometimes—with the all-consuming commitment to Mabel.

A Disappearance

When Camila began therapy with me, she said it was because she couldn't stop thinking about her father's disappearance from her life. Her father had been ill, and when she returned from a summer vacation, he was not there.

Camila's father was from another country and she had been brought up with its traditions, even though she was born in the United States. She told me her primary-care physician had urged her to try psychotherapy. She couldn't and wouldn't tell her mother, who absolutely— she emphasized the words—*would not approve.*

Her narrative wasn't long or involved. Her father had been sick when she was eight. His illness had begun when she was seven, she thought. Each day just before she left for school, she went into his room to see him. She would be dressed in her school uniform and eager for the day. After school, she would want to tell her father some of the things she'd learned that day, and when her brief afternoon visit with her father was over, she had to leave him sick in his room.

She would always remember his smile. To her, it meant how happy he was with her stellar performance in school, and how much he loved her.

The summer she turned eight, for the first time it was arranged that she would visit her father's country and relatives for six weeks. On the day she went away, before she left, she went to see her father, and they talked for a longer time than usual. She remembered that this time, her father motioned for her to sit beside him on the bed. He seemed very thin in his monogrammed pajamas. But he smiled and told her to be sure and eat some figs when they came into the outdoor market of the village. He told her how he had loved them a long time ago before he came to this country. He described them in detail: how the dark violet skins, speckled with green, surrounded the creamy flesh. The juice, he said chuckling, would drip down his chin when he ate them as a boy. He traced a line down her chin. He smiled, motioned to her to embrace and kiss him, and said goodbye.

When she came home weeks later, she wanted to tell her father about her adventures exploring in the woods and the hills, and about the figs. When the driver dropped her at the house earlier than her mother had expected, only the housekeeper was there. Camila quickly ran by her and rushed to her father's room. What she saw riveted her. Instead of finding her father where she had left him, in his bed, her father was gone, his mattress rolled up and resting against the footboard at the end of the bed. They told her later that he had died. But she didn't believe it. Even after her mother explained that they had wanted to spare her the sadness of the funeral, she still doubted.

Eventually Camila confronted her father's disappearance, and connected it with his death. She described how, for a long time, she wondered whether her father had actually left. Perhaps he was alive, and the family had made up the story that he died. It was a child's fantasy, she knew, but it was odd that it kept surfacing, as did her memory of the day in late August when she returned home after her summer with relatives in Europe. It was clear to her, all these years later, that the reason her mother and grandmother had arranged for her to go away that summer was to protect her from her father's dying, and his death. It didn't. There was no wake or funeral for her to be with him for the last time as her relatives were.

In wishing for her father's return, unrealistic as that was, Camila had fallen into ambiguous grief. Her childhood fantasy had both protected her and prevented her from admitting that her father was dead. Her grief during the times she had allowed herself to think about his death had become a chronic pain. In therapy she addressed her chronic and ambiguous griefs, which overlapped. In finally confronting her father's death, she had come to accept that her father hadn't abandoned her in life or death. His death was a separation, not an abandonment.

Sickness before Death

Katherine began therapy several months before her father died. She came in to address the trouble she was having sleeping through the night.

As her father's illness had grown worse, Katherine and her family knew that their time together was short, and that he would probably die soon. He looked slighter whenever she saw him; it seemed there was less of him each time. But he had recovered before after bouts of illness. She thought there were a lot of times, but she couldn't really remember them.

He called her before he died. She could barely hear him when he called to say goodbye.

Katherine told me that her father's death and all the funeral activities marked the beginning of her grief and the agonizing loneliness she felt afterward. There had been an earlier period, lasting several years, when she had tried very hard to feel as little as possible: when her parents were divorcing, and then later, when her father became ill. But when her father died, Katherine's grief was deeper than any emotion she ever thought she could experience.

She wondered why she had not been more prepared for his death. He'd been sick for weeks, and slowly getting worse. Because he was a veteran, an officer in the army during World War II, he received a military funeral. The muted ceremony stirred unexpected feelings and thoughts for Katherine. Her father would have been proud, she thought. Smiling and crying, she said she could almost imagine him saluting. The officers folded the flag that was draped over the coffin and handed it to his wife, who was not Katherine's mother.

These feelings of love and pride surfaced at the beginning of Katherine's grief, as she described her father's gentleness and remoteness. It seemed that her memories of him had become a touchstone for her after his divorce from her mother. She said she thought she was as proud of him as he probably was of her and her siblings. She couldn't finish her words at first, telling me how his funeral had opened up feelings of admiration and respect because of his military service. His military past wasn't anything her father wore on his sleeve, or even talked about, but she remembered the photos of him in his uniform. The funeral honored her father: "Taps" was played, and her stepmother opened her arms to accept the folded flag. She was grateful for this tribute. It awakened memories of her relationship with her father from the time before he became ill, and the time before her parents' divorce. The military funeral increased her remembered love for her father.

For a long time, Katherine's grief was forceful: she fully experienced her sorrow. She'd cry without saying anything. Sometimes she'd talk about her father or the family. We discussed how her anticipatory grief, the awareness and expectation of her father's impending death, had probably contributed to the problems she'd had sleeping. Knowing that her father was ill and seemed to be slipping away was an awareness just below the surface of her consciousness, something she did not think about while awake, but would bother her as she would start to fall asleep.

After a few months, as she began to smile more, we both knew her grief was receding. Katherine's grief was beginning to run its normal course.

Replacement Fathers

In her book *Fatherless Women: How We Change after We Lose Our Dads*, Clea Simon describes how she grieved after her father died, writing of her surprise at the emotional places where grief towed her, and how she grew while experiencing a sorrow that overwhelmed her.

At first, she became enamored of spaghetti westerns. These were not films she watched before her father died. But after her father's death, there was something about the resolute steadfastness of Clint Eastwood's distanced and deliberate characters that kept her spellbound. She knew that Eastwood portrayed an idealized, heroic male who was strong, poised, and confident: "He was unbeatable. He was never overwhelmed by plumbing or by a family. And unlike my dad, he never died."[18] Eastwood was the male ideal who grounded Simon's grief.

In a similar way, Helen Macdonald fixated on Alec Guinness, who played George Smiley in a British television drama. Just as Simon watched endless Eastwood movies, Macdonald viewed the Smiley series over and over. She imagined herself in the Whitehall offices and gentlemen's clubs where plans of espionage and betrayal unfolded. And like the action in the American western that was slow and deliberate, the British television drama for Macdonald was "glacially slow and beautiful."[19] Both Simon and Macdonald describe being captivated by the steady predictability of those heroic characters and watching them play their roles over and over, at the bidding of Simon and Macdonald. They were brave, determined men not subject to life's ultimate tragedy—death.

Staying with Father

Darian Leader, a British psychoanalyst, writes of the method a small boy used to express his anguish after his father's death.

Leader writes, "After his father died, a five-year-old boy would fit himself into a suitcase in the corner of a room, where he would remain motionless. When a friend asked his mother what he was doing, she replied he was just sitting in a suitcase. Yet, as he saw clearly many years later, he had created his own private coffin, an enclosed space where he could act out an identification with the beloved father whom he had last seen in a coffin."[20] Leader goes on to explore how the nonverbal language the boy used created a unique representation of his loss. Without speaking, the boy had created a silent tableau depicting his father motionless in a coffin—a funeral ritual in pantomime.

An adult male cousin of mine expressed a somewhat similar reaction to his own father's death, though he used words to express his feelings. My cousin was so overwrought after his father died that at the burial, he cried, "I want to throw myself into the grave with him."

Both the boy and the man expressed the pain of separation from their fathers by demonstrating their desire to remain physically with their fathers in death.

What happens to sons and daughters after being rendered fatherless? Who do we become after our fathers die? It's different for each gender, and, of course, for each person. A daughter may retreat while she grieves, or watch movies featuring indestructible heroes, or decide to raise a goshawk. A son whose relationship was problematic might displace paternal conflicts onto others in the form of disagreements, quarrels, even estrangement.

But what of the fathers who weren't there, either in person or in spirit? What grief follows the children after absent fathers die? For sons and daughters with an absent or abusive father, grief is hard to manage. The father's death may be liberating, but that freedom will be tinged with remorse and longing for what could have been.

No matter who grieves, psychological issues will surface. Issues from early developmental stages will be reactivated, whether they are attachment or Oedipal issues or later, age-appropriate reactions. Grieving shares space with the developmental issues that will be part of the psychological work and growth of sorrow. Grief can help the survivor grow.

Helen Macdonald says that grief broke who she was, and she came apart. This feeling seems nearly universal. Further, after the anguished effort of letting go of a father, grieving him whenever time and life allow, we are changed. That is part of the putting back together.

But we never entirely let go of our father: that symbolic figure with power and authority, and the personal figure who loved and taught, or who wasn't there. The living one is gone, but the internalized one, the one who becomes a memory, remains.

NINE

Children

We have lost the joy of the household.—CHARLES DARWIN

When a child dies, parents are plunged into a state of over-whelming grief. No matter the age—infant, toddler, or adult—for a child to die before a parent feels unnatural, as though the sun were unable to rise. A child dying before the parent is not how life is supposed to be.

In the past, before the use of antibiotic medicines became widespread, children's deaths were not uncommon: children died with regularity from disease. And in the past, grieving tended to be more public. Victorian parents expressed their grief outwardly by wearing appropriate mourning clothes for a designated pe-riod. During the late nineteenth and early twentieth centuries, too, works of "consolation literature" about the loss of a child became popular. Indeed, consolation writings for loss of a child had so penetrated American culture that several years *before* his own son died, the poet Eugene Field wrote his long-remembered poem "Little Boy Blue." Field's poem was published in the Chicago weekly literary journal *America* in 1888.[1] It begins:

> The little toy dog is covered with dust,
> But sturdy and staunch he stands;
> The little toy soldier is red with rust,
> And his musket molds in his hands.
> Time was when the little toy dog was new,
> And the soldier was passing fair;
> And that was the time when our Little Boy Blue
> Kissed them and put them there.[2]

The dust and the rust on the boy's toys represent the enduring nature of the grief that follows the death of a child.

Another of Field's poems that is considered a classic, "Dutch Lullaby," is also known today as "Wynken, Blynken and Nod."[3] Both poems capture how, in the nineteenth century, children were viewed by adults as both precious and fragile. In the lullaby, a drowsy child is soothed to sleep with images of sailing in a wooden shoe, a laughing moon, and a wind ruffling waves of dew. In the elegy about a boy who has left and not returned, the boy is remembered with images of his "little toy friends" who stand, "Each in the same old place—/ awaiting the touch of a little hand,/ The smile of a little face."[4]

It was also during the Victorian era, in 1851, that Charles Darwin and his wife, Emma, lost their oldest daughter, Annie, when she was ten years old. A week later, Darwin wrote a memorial letter in which he imagines what will be his and others' recollections of his daughter: "I write these few pages, as I think in after years, if we live, the impressions now put down will recall more vividly her chief characteristics. From whatever point I look back at her, the main feature in her disposition which at once rises before me is her buoyant joyousness tempered by two other characteristics, namely her sensitiveness, which might easily have been overlooked by a stranger & her strong affection."[5] Darwin's daughter Henrietta, in later writing about her sister's death, described her mother as "rarely" speaking of Annie. In describing her parents' sorrow, Henrietta stated, "it may almost be said that my mother never fully recovered from this grief," and that her father "could not bear to reopen his sorrow, and he never, to my knowledge, spoke of her."[6] Based on the correspondence in the Darwin archives at Oxford, the grief that Darwin and his wife, Emma, experienced after Annie's death continued for the rest of their lives. Despite the generous length of time thought acceptable then for Victorian parents to mourn, as well as the declarative clothing, funeral accoutrements, and extensive written correspondence that were part of grieving during that era, such unabated grief could be considered chronic.

Today, the expectations and customs for survivors of the death of a child are markedly different. Grief is expected to be briefer and more private: there are no clothes that indicate a beloved child has died. Given these changes in the culture of grief, today many nineteenth-century poems can be seen as sentimental. In their time, however, they were seen as comforting. And despite the different mourning practices of today, parental grief remains as deep and sorrowful as ever: the death of a child is devastating.

Sometimes the best way to understand the grief of those who have survived the death of a child is to hear it expressed through music. Jimmy Greene, a saxophonist, composed a moving tribute to his six-year-old daughter Ana, who was killed in an elementary-school massacre. Nate Chinen, in his *New York Times* review of Greene's poignant album *Beautiful Life*, describes the gorgeous array of singers and instrumentalists featured in Greene's work.[7] The album also includes an audio segment of Greene's daughter Ana singing "Come, Thou Almighty King," with her older brother, Isiah, accompanying her on the piano.[8] The multi-dimensionality of this luminous compilation of poetry, instrumentals, songs, and narration is intended by Greene to ease the searing pain of losing Ana, by translating it into a musical remembrance of the joy that she brought to her parents and family. Greene transformed his grief into a work of art that memorializes his beloved daughter.

Sudden and Unexpected Deaths

The death of a child is cruel, whether it's unexpected or foreseen. A child may die unpredictably, as when, in 1921, Marigold Churchill fell ill with a respiratory infection, then suddenly succumbed to septicemia. Or parents may live for months or even years knowing that their child's medical diagnosis points to a possible, or almost certain, premature death.

Some disorders, however, aren't visible until the first symptoms appear. Sukey Forbes writes of her six-year-old daughter Charlotte, the middle child of three, who had a rare genetic disorder called "nontriggered malignant hyperthermic reaction." The extended family was on vacation when her daughter first experienced symptoms of the disorder. It was Forbes's sister-in-law, Anne, a chief of pediatric anesthesia, who correctly recognized the cause and explained the condition in detail to her brother and sister-in-law. The trouble, Anne began, occurs in the skeletal muscle cells. "That's why we saw the rigidity in Charlotte's limbs last night. All it takes is a touch of fever from a cold, the flu . . . anything. Hypermetabolism moves through the body like a wrecking crew. Potassium leaks out of the cells and into the pericardium [lining around the heart] and you go into cardiac arrest." Most regrettably, there was nothing to be done. Two years later, Charlotte died in a hospital with her mother beside her. Forbes writes that Charlotte's last words were a shriek: "I want my mummy."[9]

Forbes describes her grief as unreachable, walled off by her New England Protestant upbringing. She lived stunned by Charlotte's death and by her inability to grieve. For months she shed no tears. Walking in the woods at one point, she quotes the poet Ralph Waldo Emerson, her great-great-great-grandfather: "Emerson had walked these same paths, and he had mused over similar mysteries, and, like me, he had not found it easy to emote. 'I chiefly grieve that I cannot grieve,' he wrote after the death of his son."[10]

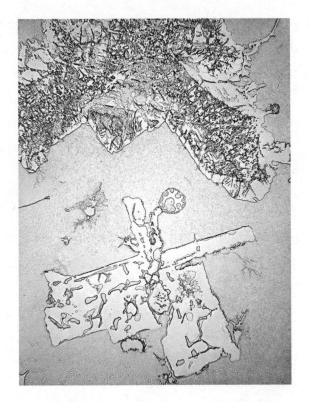

The pull between attachment and release. From *The Topography of Tears*,
© 2017 Rose-Lynn Fisher, published by Bellevue Literary Press.

It took months after her little girl died for this mother to let
go and truly cry—for her grief to be felt as a physical experience.
It finally happened during the first winter after Charlotte's death.
Forbes and her family went to Utah for a ski vacation that had
become a family tradition. On the third night of the vacation,
as she was lying in bed, Forbes says she "felt a huge sob catch in
my throat. I needed to get rid of this knot, so I forced it out, and
suddenly the floodgates just blew open and the tears came and

once they came they would not stop. I sobbed so hard for so long that my abdominal muscles got sore."[11] With the intense, wrenching sobbing, Forbes was able to *feel* her loss; her forceful weeping was her body's physical expression of her long-suppressed emotions. Psychologically for Forbes, her tears helped her to begin to grieve the loss of Charlotte, and work toward accepting her death.

Continuing Guilt and Endless Grief

What a patient of mine, Mrs. Hamilton, experienced after her thirteen-year-old was killed wasn't absent grief. It was complicated grief.

Before Mrs. Hamilton had children, she'd had a career as a financial analyst. After college she earned an MBA degree and worked for several years before she married. But when being a mother became more important than her career, she stopped working outside the home. The three Hamilton children, two girls and a boy, excelled academically. Each played a musical instrument and was involved in a sport. But it was Nicholas, the middle child, whose interest in a sport was all-consuming. His passion was archery. He read and reread books about archery, and books about archers who became heroes, such as King Arthur in *The Once and Future King*.[12]

Nick was also committed to learning how to be an archer, and his father bought him a bow that was properly sized for him.[13] Watching their son, both parents marveled at his dedication to practicing intensively as a member of his archery team.

When Nick's team entered an out-of-state field archery competition, he was determined to go. His parents agreed, and his mother accompanied the team as one of the chaperones. The team traveled to the event by bus. Nick, disappointed but happy, placed third in the competition. On the trip home, he told his mother he was going to sit with his friends in the back of the bus. It was late, almost midnight.

Months after Mrs. Hamilton began therapy, she was able to describe what happened that night. She was jolted awake when the bus entered the exit ramp much too fast, left the road and crashed down an embankment. Inside the bus, amid the darkness and with luggage strewn about, Mrs. Hamilton tried to get to the back of the bus to Nick, but couldn't. There were EMTs who helped her out of the bus, then told her to wait. She gave the police all the identifying information. Her husband met her at the hospital, and as they were shown into a small room, Mrs. Hamilton told him she hadn't been able to find Nick. A doctor came in and sat down with them. Gently, he asked them if they would go with him to identify who might be their son.

Nick hadn't survived the crash. Sitting in the back of the bus, he and his two friends had been killed by the impact of the bus rolling down the embankment. When they went in to identify Nick, Mrs. Hamilton said her husband held her around her waist to steady her the whole time they were in the room with Nick's body.

In our first sessions, Mrs. Hamilton would talk only about her daughters, what they were doing in school, and often what she'd cooked for dinner. She did comment, offhandedly at first, that she hadn't changed anything in Nick's room, and didn't want anyone to go in there. The only difference in the room was the recovered backpack, which she'd placed on the floor next to the desk where her son usually dropped it. The bedroom door stayed closed and the room was off limits to the family. But sometimes, she said, she opened the door to Nick's room when her husband was at work and the girls were at school. She said she'd usually just stand there. But sometimes she'd walk in and look at Nick's things. Like the toy soldiers loved and left by "Little Boy Blue" in Eugene Field's poem, Mrs. Hamilton knew how much the archery books in his bookcase had meant to Nick, and now to her.

After months of the room being closed off to everyone most of the time, Mrs. Hamilton agreed to try psychotherapy. She called me after a friend gave her my name. We met for many sessions before this bereaved mother could talk about her son. My own silent tears began before hers flowed. Later, my tears would often mirror hers. She began to talk about Nick, the young archer, her only son who was so enthusiastic and so honest. And she cried.

Mrs. Hamilton suffered from a complex set of emotions that, in turn, had complicated her grief. She had been traumatized by the bus rollover and crash. And she was experiencing survivor's guilt. She had gone on the trip to chaperone the team, and at some level to watch out for her son. Her grief, her trauma, and her guilt were incalculable weights. It hurt her heart, she would whisper, touching the left side of her chest. And it was the guilt, which overrode her grief, that she needed to express. There were many times she'd talk about how, as one of the chaperones, she was there to be sure that nothing happened to the kids, and to Nick.

The worst that could have happened, did happen. Nick and two other boys were killed in the accident, and my patient survived. Whenever she spoke of the trauma of the crash, it was always in terms of Nick, not herself. At some point she was able to describe her experience of being awakened by the bus rolling over, waking her, and her fear about her son. And the traumatic aftermath of not being able to reach Nick.

There were many sessions when all Mrs. Hamilton could focus on was how she had failed to protect Nick, and the guilt she felt because she had survived the crash, while her son and the other boys did not. As the enormity of her guilt lessened, her grief increased, slowly and fitfully, but surely. She learned words that belonged to her grief: nouns, adjectives, and verbs that made up the world she had been unable to be a part of for a long time. She started using words referring to Nick that were grief words. She said that finally her thoughts about her son were tethered in the past. Her son Nick had died.

She described how she loved Nick, how different a son was, and how she missed him. She had begun therapy with the story of Nick, her son who loved archery, and it became a continuing narrative of this lost teenager whose charm and strength could never be forgotten. As this bereaved mother finally began to grieve, she acknowledged her trauma from that fateful night, and her guilt.

The Anguish of Self-Inflicted Deaths

There are other types of unexpected deaths, not only ones that result from vehicular accidents like Nick's. Some of these youths, and adults not so long out of childhood, die accidentally from a drug overdose or other reckless decision. And some die by a deliberate act of self-destruction. These self-inflicted deaths by children can shatter parents with guilt, anger, grief, and questions.

Oskar Eustis, parent and artistic director of the Public Theater in New York City, was directing Shakespeare's *Hamlet* in the summer of 2016. Michael Paulson writes that as Eustis watched the play, he shed tears "when the ghost of the dead king cries out to his son, Hamlet; tears when Hamlet laments that he has 'lost all my mirth'; tears through 'To be or not to be,' the speech in which Hamlet considers ending his life." Eustis's son had taken "his own life nearly two years" earlier. In his article, Paulson reflects on "the possible connection between the playwright's grief and his play," and the likelihood that the famous tragedy "had new and unwanted resonance" for Eustis. Later in the article, we learn of Eustis's conviction that art must be true, but in so much of what's being produced, the works "are lying to the audience for the sole purpose of making the audience feel good." In his view, "that's not what art is for. . . . The mission that I feel like I have is to figure out how you can tell the truth about how tragic and unfair life actually is without destroying hope."[14]

Parents who are confronted with a son or daughter's suicide may be stunned by the initial anger they experience, in addition to the guilt. It's important to understand that these emotions are not unusual. Indeed, they are part of what may be the most difficult type of bereavement to be experienced by a survivor.[15] By confronting their anger, guilt, and questions after a child's suicide, and by allowing these emotions to be felt and voiced along with the healing of tears, parents are able to grieve.

When Adult Children Die

When a child grows into an adult son or daughter, one who has been enjoyed and appreciated by the child's parents—when this child, now the adult, dies—what kind of grief is that? Staggering news hits the parents. It flings them across the room of their lives with an impact that some cannot recover from, while others rage against recovery.

It happened to Roger Rosenblatt, the author of *Making Toast* introduced in Chapter 7, when his daughter died. Grief crashed into him when his daughter Amy, a pediatrician, died at the age of thirty-eight, collapsing on a treadmill in the family's downstairs playroom.

Amy and her husband, Harris, a hand surgeon, had three children. The youngest, "Bubbies," was a little over a year old when Amy died; the second, Sammy, was five; and the oldest was Jessie, seven. Sammy and Jessie discovered their mother on the treadmill. Jessie ran upstairs to Harris saying, "Mommy isn't talking." Despite Harris getting to Amy in seconds and administering

CPR, there was no saving her: her heart had stopped. Amy's sudden death was attributed "to an anomalous right coronary artery."[16] This congenital cardiac condition can exist without symptoms, and often isn't recognized until an initial event that can cause heart failure and death.[17] In Amy's case, her two arteries were not on separate sides of her heart, but instead "ran alongside each other," and "could have been squeezed between the aorta and the pulmonary artery, which can expand during physical exercise. The blood flow was cut off . . . she could have died at any time in her life."[18]

Rosenblatt's grief came in enraged surges. He tells of the shattering effect his daughter's death had on him, and of the profound changes that Amy's death caused in his and his wife's lives. The day they heard of Amy's death, they drove from Quahog, a summer village on Long Island, New York, to Bethesda, Maryland, where the family lived, now without a wife and a mother. When Jessie, the oldest grandchild, asked Rosenblatt how long they'd be staying, he answered, "Forever."[19]

As mentioned earlier, artists use many genres and media to eulogize, interrogate, articulate, and rail at the loss of a child, no matter the cause of death. The book-length poem *Gabriel*, by the poet Edward Hirsch, is a testament to a father's love and grief. Hirsch's adopted son, Gabriel, died of cardiac arrest at age twenty-two due to a drug he used at a party. Hirsch's poem rings with his suffering.[20] Memoirist Emily Rapp, in her review of Hirsch's poetic eulogy, expresses her own grief at the loss of a child. Around her review of Hirsch's poem, Rapp envelops a description of the anticipatory grief she feels due to the impending death of her one-year-old son, Ronan, who had Tay-Sachs disease (he later died when almost three years old). In her review she writes, "The work of grieving, Hirsch rightly suggests, never ends. . . . While 'Gabriel' makes 'the chill grip of grief' no less despairing, Hirsch's fierce grappling with it makes the world less lonely for those of us who still fear that daily climb into our lives where a beloved child no longer exists. There is some consolation in this, no matter how grossly inadequate."[21]

When an elder adult is confronted with the death of an adult child, the news of this devastating blow can be met with sustained disbelief. In a *New York Times* article, Paula Span describes the lingering trauma that an elder woman experienced when informed that her adult son had died.[22] Span writes that when Mrs. Giotta, eighty-seven years old, was told by two police officers that her fifty-one-year-old son, Michael, had been found dead in his home, apparently of a heart infection, Mrs. Giotta didn't believe them, saying: "Don't tell me lies like that." This mother, who had come to the United States from Ireland sixty years earlier, couldn't hear what she had thought to be impossible.

Mrs. Giotta's daughter, Kathleen, said her brother Michael was the one in the family they would contact to be sure their mother was all right. "He was the one the rest of us relied on," she said. "If Mommy's not answering the phone, who do we call? Michael, who lived five minutes away." When an adult child who has provided emotional and practical support to a parent dies, the caregiving functions they provided come to a halt, which compounds the burden of grief. As Mrs. Giotta explained, "You lose a part of yourself." Marsha Mailick, a social scientist who has studied bereavement, is quoted by Span as saying that the death of an adult child "is a trauma that doesn't go away."

Indeed, such grief can damage the parent to the extent of shortening their own life. When a friend of mine and I once were talking about the death of adult children, he mentioned an acquaintance of his, a mother, who had suffered terribly after the death of her adult son. He told me, "She died within a year, because, you know, the heart breaks."

Falling Out of Time, David Grossman's "strange and riveting book," is "partly a folk tale, partly a play, partly a novel in verse. There's no genre to describe it."[23] In prose, Grossman renders a son's death powerfully and intimately. Even five years later, both parents are still struggling with the memory of hearing of their son's death. The bereaved father, who represents every bereaved father, continues to be stricken by his son's death, and even seems at times to be distanced from reality. One day, at the dinner table, he tells his wife he has to go. "Where?" she asks. "To him . . . there," he replies. "What do you mean, there?" she asks him. "I don't know," he replies, but the very act of telling his wife that he wants to go "there" shifts his grief. He hears, through the voice of a boy, "There is / Breath / there is breath / inside the pain." And he begins to recognize that words hold truth, and the truth is that "the boy is dead." The man has found the words that belong to grief's language. It is a language unto itself, and those words are ones that still break his heart. But now, after "five years on the gallows of death," he has found the way to grief—he has faced the reality that his son has died.[24]

In Grossman's book, the man talks, and he walks, and he meets other parents who have experienced the heartbreak of their child's death. The book ends with all of them stitched together by an imaginary thread that reaches from the place of "there" to the place where they all met, where they became stitched one to another in their shared grief.

Lingering Grief

Ms. K began therapy soon after she retired. In describing her history, she told me that years earlier she and her husband had their first child, a girl born with a damaged heart. Baby Angela had undergone several surgeries, but they were unsuccessful, and she had died before she was a year old. Two years later, Ms. K had a son who grew up to be healthy and strong. Ms. K, who was a computer analyst, went back to work after Henry began kindergarten, and worked until she retired, which was several months before I first saw her.

Retirement changed things for Ms. K. There was time to think, to feel, and to remember. The anniversary of the death of Angela coincided with the month she retired. At first when Ms. K became upset, she thought it was because she'd lost her professional identity when she retired. She decided she was depressed. She realized that without going into the office each day, there was a lot of time for emotion. She said she was experiencing emotions she hadn't felt before and didn't know where they were coming from. That was why she had decided to begin therapy.

During our first session, a consultation, Ms. K said that she'd had two children, but only one lived. She'd had a baby girl who had died when she was nine months old. She had a son several years later. Henry had grown up to be intellectually curious and driven. Now in his twenties, he was in his third year of a doctoral program in chemistry. The focus of therapy quickly came to settle on the baby Ms. K had lost. It was clear to both of us early in our sessions that there were emotions around Angela's death that Ms. K hadn't fully experienced. She said she'd been busy raising Henry and working full time. Her early grief was initially felt, but suppressed given the pace of her life. And she discovered that the trauma of the surgeries her baby went through had added to her grief.

Ms. K recounted painful details of her baby's brief life: how much of it was spent in the hospital, and how often she and her husband were at the hospital with Angela. She described a tortuous series of cardiac surgeries that couldn't repair Angela's heart. And Ms. K remembered being awakened at 2:30 a.m. by the penetrating ring of the telephone. It was a call from the hospital, and she listened to someone telling her that her baby had died. She said she had screamed, "No," into the receiver. After quickly throwing on some clothes, she and her husband drove to the hospital to be with their baby for the final time.

Although she'd mourned in the past, it was not until one of

our therapy sessions more than twenty years later that she found the freedom to find and express her sorrow, which had been cut short when she went back to work soon after her baby's death. She cried, speaking words she'd never spoken before, saying she knew what it was like to hold a dead baby.

Therapy became a haven for her to feel and express grief that had been delayed for years. Now she had the time to confront what she couldn't before: the trauma of the surgeries, rocking her baby in the hospital, and the phone call announcing her baby's death. There wasn't time after Angela died to feel her grief fully and to let it run its course. Having another child soon after had renewed and diverted her energy, she said, and she had thrown herself into being a mother and a professional. But now, in therapy, there was a place where what had been unreachable could come forward with a voice. It was safe, and there was room to revisit feelings about her baby whose cardiac problems were so severe that she could not live.

The pain was as strong after she retired as it had been years earlier, during the medical efforts to repair Angela's heart, and upon Angela's death. This surprised her. But we talked about how her retirement had finally given her the emotional space and time she needed to access and experience her grief, which had been delayed for so long.

The aching chasm that comes after the death: how empty the world feels without the child, however young or old. What was familiar about the child—what made lips smile, and eyes light up—and what was anticipated for the child-to-be, are gone. How strange is the silence that echoes with the memory and the anticipation of a voice. Now there is no planning for this child, who has no future. Grief takes over the brain, the heart, and the body of the parents as they move into a world reshaped and strange: a world without their child.

TEN

Sisters and Brothers

My grief lies all within, and these external manners
Of lament are merely shadows to the unseen grief.
—WILLIAM SHAKESPEARE

The grief of losing a brother or sister is the least acknowledged
of bereavements within the immediate family. For how can the
grief of a sibling take center stage? The parents own this grief—
it belongs to them.

Beyond the family, too, sibling grief is often overlooked, or
even ignored. Friends and acquaintances will express sympathy
to the bereaved sibling, then quickly shift their attention to the
primary grievers, the mother and father. Condolence-givers will
say, "How are your parents holding up? You must be a big help
to them." The ruin in which the death of a child leaves parents
can push aside the grief of their surviving children. A pattern

becomes entrenched for the living sibling—their grief, unrecognized, stays outside the circle of parental sorrow. Kenneth Doka has called sibling grief "disenfranchised" because siblings' claim to grief over the death of a brother or sister receives scant attention.[1]

But a sibling feels intense pain when their sibling bond is broken. They suffer the absence, the ache of missing the sister or the brother who is no longer there. Siblings create a whole private world of shared memories as they grow. Sharing their experiences in endless ways, siblings play together, fight, work, tumble, cry, and laugh together. "Not another peep!" a granduncle who is babysitting calls out to his chattering nieces, whom he had tucked into bed an hour before. The granduncle hears giggling, and then: "Peep, peep." To share genes and to grow up together—from toddlers, to children, to young adults about to leave the family—belongs exclusively to a sibship. To live the vagaries and joys with one another makes the filial bond tight and lasting, a thing to cherish and draw strength from.

Siblings will compete, either destructively or with healthy vigor. They vie for parents' attention and approval, and fight to top each other. In sports, siblings outperform and lose to one another. Parents, aware of the effects of positive and negative competition, can steer these rivalries into healthy channels. Brothers and sisters clamor to be the smarter student or the better athlete, sometimes at the expense of a sibling and sometimes with the support of a sibling, as the dynamics of growth propel young siblings into teenagerhood, where development continues to sort them out. After wriggling in and out of different personas, they become, as all adults do, different from who they were when they were childhood siblings. Nevertheless, to lose that playmate, friend, and confidant—to lose the history they lived together, and all that they were to one another—can leave the bereaved sibling detached and adrift.

Sibling Genetics

Siblings share about 50 percent of their genes with each other, and all of their genes come from the same parents. It is this inherited closeness, this genetic, filial connection, that for centuries has informed the expression "Blood is thicker than water." Genetic similarity can serve siblings in many ways: in a medical situation, for example, when there is desperate need for a replacement organ. The opposite of "an eye for an eye" taken in retribution, a kidney may be given for a kidney in magnanimity—to save the life of a brother or a sister.

Sibling grief, faint at first, can emerge years later when something, some occasion, pushes through a psychological barrier erected long before.

A friend called me one morning, distressed after returning from her annual summer vacation with her family. She was still upset about something that had happened when everyone was getting ready to leave the beach house. Her car was almost packed, and she was helping her mother get her things together. After taking a last look around, my friend and her mother stood in front of the fireplace, chatting. Somehow there was a segue that brought her and her mother to the topic of her younger sister's death. My friend said she decided to use the opening to try to describe her feelings about the death to her mother—emotions that she had trouble finding words for even after hearing all her life about her younger sister, who had died as a toddler.

Everyone in the house heard her mother's response. It was swift, angry, and tearful. "That was *my* pain, not yours! It was *my* little girl who died."

For the mother and daughter at that point, both ready to say goodbye and with other family members in tow, clearly the unexpected exchange had to stop. My friend said that it seemed as though her mother was keeping a tight grip on her grief. "My mother owns it, she always has, and can't recognize that all these years I've had my own grief of never having or knowing a sister." Even so, my friend told me that she was determined to pursue the subject with her mother at some point. She thought it was important for both of them to recognize her grief at the loss of a sister who never grew up with her.

Disenfranchised grief, denied a legitimacy to be recognized, can haunt siblings, lingering for years in the limbo of disapproval and adding dimension to the loss. The marginalized grief that siblings experience when their loss is unrecognized or rebuffed may be displaced and later experienced in other relationships.[2] When a loss isn't considered sanctioned—it's not approved of based on a family's mores, or on society's—the lack of recognition will add to the anguish of disenfranchised grief.[3]

A Forbidden and Misunderstood Grief

In addition to being forgotten and disenfranchised, grief that siblings experience may be forbidden. If a particular death is a topic not to be spoken of, then the grief that follows is likewise not to be expressed. Parents may even explicitly tell a child not to speak of their sibling's death.

The difference between what the primary bereaved, the mother and the father, suffer and what the secondary bereaved, the sibling, suffers can be stark. The biological relationship, and its meaning in familial and social terms, is what contributes to the difference between primary and secondary grief. The pain of the primary grievers encompasses their loss of all that could have been—the parents' future with and for their child, and their hopes and expectations for continuing the history of their family. Yet secondary grief—the sorrow that is disenfranchised, marginalized, forgotten, and sometimes even forbidden—can also cause great pain, and should not be overlooked.

Research on sibling grief, like the attention we tend to give siblings who are grieving, is hard to come by. As researchers David Kissane and Nadine Kasparian observe, "What does become apparent as one spans the literature about sibling bereavement is the relative dearth of studies exploring this field."[4] That there is a shortage of research on children, teenagers, and adults who have lost a sibling is understandable, because there are so many variables to consider—including birth order, gender, ages of siblings, the nature of relationships (with parents and with each other), and the nature of the death. There are at least six different sibling pairs (setting aside other siblings, and adopted siblings, if any): sister and sister; sister and brother; brother and brother; identical twin sisters; identical twin brothers; and fraternal twins who are sister and brother. Taken together, these complicated variables present knotty challenges for researchers. Yet this neglected category of bereavement needs more investigative focus. Sibling grief matters. To lose a sibling affects the survivor physically, emotionally, cognitively, socially, and developmentally.

The many variables that affect sibling relationships point to the complexity of this relationship. It can be close, competitive, or adversarial. It can be plural: there may be several siblings. There may be just one parent, or two, or more when stepparents are taken into account. There may be same-sex parents. The family dynamics may involve a collusion with one parent against the other, or a collusion between or among siblings against their parents.

Yet however these relationships with parents shake out, the lives siblings share with each other exist in a world of their own, apart from parents. As often as siblings are allies, friends, or helpers of one another, they can also be the opposite: rivals, antagonists, even adversaries. When a family is divided against itself—by competition, envy or greed, or a history of parental dysfunction—sibling interactions can be nightmarish and siblings' lives marginalized. Choice of lifestyle, profession, or partner can further complicate sibling relationships. One sibling may choose a profession that the parents and society view favorably, while another sibling may select a career or lifestyle that is frowned on by the family. These differences can begin or exacerbate sibling rivalries. Yet even in a family where the relationship dynamics are negative and destructive, and where feelings of anger and animosity are mixed with familiarity and love, the death of a sister or brother brings an end to a significant part of the survivor's world.

After a Brother or Sister Dies

In whatever way a brother or sister grieves the loss of a sibling, each experiences a grief that's outside the mainstream of primary bereavement, and a grief that is influenced by the gender of the siblings involved. Again, it's no wonder that sibling loss has been given short shrift in research and the arts: each sibling and gender iteration carries its own signature in how the relationship was established and grew.

It is also not surprising that bereaved siblings often endure great loneliness as they suffer their unacknowledged sorrow. And what about other emotions? What about the envy or anger the grief-stricken sibling once felt toward that lost brother or sister? And what about the times when they were mean to their sibling, or at some point even wished them dead? The guilt they feel can be acute.[5] Charles Darwin's daughter Henrietta, eight years old when her sister Anne died, was distressed, telling her mother: "Mamma . . . I used to be very unkind to Annie."[6] These emotions, unacceptable to the surviving sibling, can quickly lead to a heavy sense of guilt.

And what effect will the bereaved parents' grief have on the surviving child? The survivor may be resented by a parent or parents for having survived, a situation dramatized by Judith Guest's 1976 novel *Ordinary People*, which inspired an award-winning film. Or the surviving sibling may be smothered by parents who assuage their guilt and worry by becoming overly protective of any remaining children. No matter how the mother or father decides to interact with a surviving daughter or son, that decision will have a significant effect on that child's development. And should the surviving child have to assume a caretaker role with the parents, that child may feel the need to suppress their own grief further. A survivor's shifting relationships with his or her grieving parents can generate a confusing array of feelings and obligations that can further obscure grief over the loss.

The Sibling Relationship

There are emotions and experiences that exist only between siblings, unknown to parents or other adults. In her book *The Angel in My Pocket*, Sukey Forbes writes about her family's experience of discovering that her daughter Charlotte had a genetic illness, and of her death soon afterward. Forbes poignantly shares her young son Cabot's response to being told his younger sister has just died. He shouts, "You're lying. That didn't happen!" For him, there had been no warning. Cabot is seven, one year older than his sister. When Forbes puts him to bed that night, he asks,

"Who will I play all our Charlotte games with?" His mother answers that he can play them with her, but Cabot rejects this idea, crying, "No. You don't know them, and I can't teach you because they were our secret."[7] That's a seven-year-old's definition of what it means to have a sibling, and for that sibling to die.

Children's level of understanding about death and grief depends on their age. Young children have difficulty grasping what it means for someone to die, particularly if open communication is not the family norm. Children often use magical thinking to explain the world, and may feel responsible for their sibling's death as they recall their sibling spats. When parents, because of the devastating effect on them of the loss of their child, are unable to clarify this mistaken thinking, other support systems can help in dispelling these thoughts.

For adults who lose a sister or brother, their sibling history and present circumstances affect their grief and the direction it takes. When my grandmother lost her younger sister, I was little, six or seven years old, but my memory of her reaction is still vivid. My mother and I were scheduled to take a train trip across the country, and my grandparents, with whom we were living at the time, planned to see us off. It was early in the evening, the night before our trip. We were eating dinner when we heard a knock at the door. My grandmother went to answer it, and moments later she returned and handed a Western Union envelope to my grandfather. He opened the yellow envelope and read aloud the telegram's staccato message:

"My wife Anna died STOP. Funeral in two days STOP."

"It's impossible!" my grandmother exclaimed, then burst out with a torrent of words, reminding us all that her sister, who'd visited us only days before, looked "happy and healthy." My grandmother's sister, Anna, had stayed with us for several weeks. She lived in a mining town miles away in another state, with her husband. He had sent the telegram.

My grandmother and her sister, and the two men they married, were barely out of their teens when they had come to the United States. My grandmother arrived first and married my grandfather—later they had children and then grandchildren. My grandaunt was my grandmother's beloved younger sister. Her immigration had been sponsored by my grandparents, who took her in to live with them after she arrived in the States. They found a job for her in the mill where they both worked, my grandmother on the looms and my grandfather as the master dyer. Anna was a beauty, with a radiant personality that everyone loved, and she hated working in the mill. Eventually she married a man no one else liked and moved with him to the mining town where he worked. They had no children. Every summer my grandaunt came to visit, and we would be delighted and entertained by her sparkling personality, her warm attention, and the presents she brought us.

For my grandmother, the blow of her sister's death, so unbelievable at first, seemed intolerable. She had promised her parents that she would watch out for her younger sister after she came to this country. My grandmother's sorrow took on the weight of the grief she imagined her parents would have felt, as well as her own grief, that of a supportive and close older sister. No longer would her sister be the charismatic relative who visited each summer and charmed my grandparents, other relatives, and me. No longer would there be those times when they traveled back into their memories, regaling one another with stories of their Mama and Tata while they laughed, cried, and felt rejuvenated with joy and sadness in their shared reminiscences. The *joie de vivre* that my grandaunt brought to the household—that would never happen again.

That evening, my grandmother's cries and muffled wails of grief were heard throughout the house. My grandparents weren't able to come with us to the station the next morning to see my mother and me off on our trip. That morning my grandmother's tears and mine flowed: mine for leaving my grandmother, and hers for losing her vibrant sister.

My grandmother's connection to her family, and to her mother country, ended with the death of her younger sibling. Her sororal grief was not secondary or unrecognized: this grief belonged to her, and it was primary because it included the parental proxy she carried for her mother and father.

The Ghost of Siblings

When the grief of a parent lingers for years after the loss of a child, the survivor child or children can feel the "ghost" or presence of their deceased sibling as they grow up. Charles Darwin was never able to speak of his daughter Annie after she died, and Henrietta, Annie's sister, writes of the unending sorrow her mother felt.[8] Emma Darwin's grief after Annie died, as described by Henrietta, reminds me of the chronic and abiding grief my own mother felt after her infant daughter, my sister, died. Not until I was an adult did I come to understand that my mother's sadness was her way of carrying my sister's death with her until she herself died.

My mother had a sister seven years younger, the youngest child in the family, whom my mother took care of from the time her sister was a baby. Growing up, their bond was the closest of any in the family. Sepia photos captured by an old box camera show them arm in arm, smiling with carefree expressions. When my mother's beloved sister died, several years earlier than my mother, I remember that my mother was immobilized with grief. She couldn't bring herself to go to the wake, or to join the family at the burial.

Both my mother and my grandmother lost younger sisters for whom they'd taken on a maternal mantle. My grandmother traveled out of state to attend her sister's funeral services. But for my mother, it was different. Her chronic grief over the death of her daughter, my sister, had become a continuing undercurrent of sadness. Years later, to lose a beloved sister she'd helped to raise had the effect of landing another maternal blow. There was no emotional reserve to enable her to take part in the funeral rites for her sister, the rituals that make a death real to the survivors. Her maternal grief was simply too heavy.

Never Found

Grief follows the loss of an adult sibling who goes missing just as surely as grief follows news of a death that arrives by telegram. Such a catastrophe happened to a patient of mine, Brad, who suffered years of ambiguous grief after his younger brother, Dan, vanished while hiking.

Several weeks after Dan's disappearance, Brad traveled to North Carolina to look for his brother. When Brad set out, he felt no grief at first, only determination to find Dan, mixed with concern and a good deal of hope. Alarm took over, however, not long after he arrived in Hot Springs, where some calls from residents suggested that Dan may have been sighted. There is something about a window of opportunity, he told me, that shrinks as the days stack up after a person vanishes. Time pressed down on him at first, he said. Then it sped up. Then it stopped. He lost a lot of the hope he started with, and fear settled in the pit of his stomach.

Brad's search was dogged and led to anyone with knowledge of the stretch of Appalachian Trail between Hot Springs, North Carolina, and Damascus, Virginia. He hired a man who had hiked and worked on the trail for twenty years, and who guided Brad to trail shelters and small towns adjoining the trail, looking for signs of his missing brother.

Of the three brothers in the family, Brad, twenty-five, was the middle one. His older brother, Roger, twenty-nine, was called "the professor" by family and friends because he knew the answer to everything. " 'Just ask Roger, he'll know,' was the caption in his high school yearbook," Brad told me. Dan was the youngest, an athlete, hiker, and climber who at twenty-three had decided to take time off from graduate school "to see parts of this country few people have ever seen."

As he searched for Dan, Brad said, his emotions kept bumping into each other, triggering agitated thoughts that sloshed around in his head like clothes in a washing machine, back and forth, back and forth. Alarm, anger, and determination mixed with happy childhood memories, like when he and Dan played hide and seek in the woods. These memories and feelings kept cropping up and competing with one another. Often, they interfered with sleep at the end of the day, though when he was weak from fatigue, and raw from hiking and climbing, he would fall into a fitful sleep. No matter where he and his guide spent the night, the bedding was always hard.

Dealing at times with emotions other than grief was the hardest. Dan's plan had been to hitchhike from where they lived in New England down to North Carolina and join the Appalachian Trail at Hot Springs. From there, Dan figured it would be about a two-week hike north to Damascus, Virginia. Brad told me it was just like Dan to do something like that: he was physically fit, strong, and adventurous. Brad felt anger at his naïve younger brother, who had trusted that everything would be fine. But that wasn't all: in addition to uncertain weather, snakes and bears in the wilderness, and Dan's unfamiliarity with the terrain, he resisted carrying a compass. Brad had given him some detailed maps of the trail and bought him an expensive compass, but he found the maps and compass in a drawer in his brother's desk weeks after Dan disappeared.

How could Dan have put himself at such risk—why had he gone off to distant places alone? What a stupid decision!

But at other times, guilt obscured all Brad's other feelings. It reminded him of how airlines blacked out days when you couldn't use your frequent-flyer miles. Guilt was like that. Guilt that he hadn't talked his brother out of this adventure. Why, simply pointing out the dangers of hitchhiking ought to have been enough to dissuade Dan from attempting the trip at all. So when guilt kicked in, anger, hope, and grief went missing—like his brother. This was the beginning of the anguish of invisible grief, which had nothing to hold onto. Later he learned that what he was feeling was called ambiguous grief.

Days stretched on as Brad and his guide tracked down all the leads they could uncover, talking with people along the trail, reading entries in the logbooks at trail shelters, and checking in with government officials. No organized search was launched because there was no solid evidence that Dan had even reached Hot Springs and begun his journey north along the trail.

The time to leave closed in like fog, slowly at first, and then suddenly it was time to go home and resume his own life. Brad said he had felt snatches of despair. Grief was settling into his consciousness.

Growing up, Brad and Dan had been close, and probably became closer because of their father's long hours away at work.

They spent days exploring the woods with each other, even in winter. They played sports together, mostly unorganized baseball and football. Hurt feelings at times erupted because Brad, the older and bigger of the brothers, tended to do everything better than Dan. But the personalities of the two seemed to take competition in stride. They had good times together. They would bond even more when one of them broke the rules and escaped their parents' disapproval. Mostly it was Dan who would be provocative with their parents, and Brad who helped get him out of trouble. They breathed more easily after the upset subsided.

Dan had called his brother from a motel off I-81 as he was hitchhiking south to begin his adventure. That was the last time his family heard from him.

Brad's torment became galvanized whenever, unwittingly, he would imagine what could have happened to his brother. Was Dan robbed, was he killed? Did he lose the trail in a storm? Did he fall, injuring himself, and die slowly and in pain? Death would have been quick if he had fallen from a high place. Brad was agonized by not knowing. The pain of not finding his brother, and of not rescuing him as he had done so often when they were kids, plagued him. He was the big brother who had let Dan down.

The ambiguity of Brad's grief made him feel helpless and guilty—feelings that at times were intolerable at other times were helped by thinking that Dan might still turn up at some point. Over the years, Brad's grief, and longing for his brother, intensified at anniversary times: late spring in the United States, the season when he had traveled to North Carolina to look for

Dan. His grief took years to change and recede. When finally it did, it was after the entire family agreed to hold a memorial service for Dan. Five years after Dan went missing, Brad and his family held a service without Dan's body to inter. They had decided that Dan most likely had died on the Appalachian Trail and would never be found.

With ambiguous grief there is no final goodbye, no burial. As Pauline Boss and Patty Wetterling have put it, "An ambiguous loss is a loss that remains unclear and thus defies closure."[9] Ambiguous grief clutches at scraps of hope like dry leaves, fluttering, cling to twigs through a relentless winter.

Grief on Hold

Another patient of mine suffered a sibling loss similar to Brad's, but with a different outcome. Jenene decided to begin therapy a year after her brother's body was found along a bank of the river that divided the town where she lived. She was five years older than her brother, and they were never especially close. The age span separated them, and she didn't particularly like sports though her brother, Hector, did: he loved team sports like soccer, basketball, and baseball. Still, both she and her brother were close to their parents, and the family was a tight-knit unit, going on annual camping trips that they all enjoyed.

The March weather had turned stormy on the last evening that Jenene and her parents were with Hector. She remembered that her brother said he'd be home around ten, later than usual because his basketball team was practicing long hours for a late-season tournament. But Hector wasn't home by ten. When midnight came and went, her parents called the police to report that their son hadn't returned from basketball practice. Over the following hours, and then days, everyone did everything that has to be done when a teenage boy becomes a missing person.

Hector was missing for only a few days. He liked to walk on the edges of the struts and railings of the bridge that spanned the river. Jenene always held her breath whenever she was with him and would watch him. She kept telling him it was too dangerous, but she never told her parents about it—until he disappeared. She told her parents and then the police about it, adding that her brother thought it was important to repeat the feat the same way each time he performed it. If he had fallen off the bridge, it was bad timing: more rain than usual had fallen that spring, making the river very high, almost overflowing, and chunks of broken ice had made the cold, deep, fast-moving waters especially dangerous. Hector's body was found on the riverbank several days later.

Life changed. Just as the family covered their mirrors, sorrow blanketed the household. But it was Jenene's parents who were the focus during this time of mourning. Friends spoke comforting words to them, sat with them, and cried with them. Visitors from the community brought food and sympathy to her parents as they all sat shiva. There was little opportunity for Jenene to talk about how devastated she felt. No one seemed to pay attention to the fact that she had lost her brother, her only sibling. Even though the two hadn't been close, they had shared experiences, shared secrets. Most of her parents' friends complimented Jenene on how she was a support to her parents. She was too embarrassed to express her sadness and grief. She said she felt it wouldn't be fair to her parents, that there was no room for her to be upset in the midst of their anguish.

Just days later, her mother seemed to become more protective of Jenene than she had been earlier, and as the weeks went by, her concern for Jenene's safety continued to increase. Her father, too, was shaken and distraught by losing his only son. During the days and nights that Hector was missing, her father said very little. He had survived a cruel life before emigrating from Europe to the United States. It was her father's stories that conveyed the family's history and demanded attention in the family. And now he seemed unable to tolerate the loss of his only son, who would have carried on the patrilineal line. Jenene began to feel smothered and ignored. When she graduated from the local community college later that spring, she decided to move to the city, where she started a job and rented an apartment with two friends.

Even after Janene moved out to be on her own, she said she felt as though she was holding in her grief, that she wasn't permitted to express it. We talked about what siblings often experience in a family, how theirs is a form of grief often relegated to the very edges of parental grief. Gaining an understanding that she was experiencing a disenfranchised form of grief, and putting words to what had happened, gave Janene permission to feel how much she missed her brother, and gave her the freedom she needed to grieve him.

In moving away from her family, Janene began to live her new life: outside of her family, but without her brother.

The Loss of a Twin

Twins share a sibling relationship tighter than most. There are two types of twins, fraternal and identical. Fraternal (or dizygotic) twins share 50 percent of their genes with each other, while identical (or monozygotic) twins share all of their genes with each other. During fertilization, the female egg and a male sperm fuse to form a single-cell zygote. Identical twins happen when this zygote divides, resulting in two identical cells. This rare event happens in about four of every thousand pregnancies. Fraternal twins, which result from two eggs being fertilized by two sperm, are also unusual, occurring in about twenty-three of every thousand pregnancies.[10]

Research has begun to show that the DNA of "identical twins" may not, in fact, be identical. Each time a cell divides, there's a possibility that a gene will be changed or mutated in one twin but not the other. At the time of birth, and as twins grow into adults, one twin can have a different number of copies of the same gene than the other twin. In this genetic circumstance, called "copy number variants," one twin could have six copies of a gene while the other could have only two copies, or none. Such developmental differences may be inconsequential, or not. For example, a difference might affect one twin's vulnerability to contracting a particular disease, such as arthritis or lymphoma.

Because of this genetic finding, researchers have suggested changing the term "identical twins" to "one-egg twins."[11]

Research on twin loss is even sparser than research on sibling loss. What of a twin who loses a twin? The death of a twin who has been raised with the co-twin ends a relationship that began in the womb and was lived closely together, with understandings and rivalries that can come only with being the same age, in the same family. The weight of this loss for fraternal—dizygotic—twins is intense, representing a most profound sorrow. With monozygotic twins, who initially share identical genes because one zygote split into two, the loss may be felt even more deeply.[12]

Even when one twin dies before the other can be consciously aware of the absence, an elusive ache can remain. Helen Macdonald knew that she, tiny and premature, was in an incubator after she was born and before her parents took her home. What for many years she didn't know was that she had a fraternal twin, a brother who was born with her, but who died soon after birth. Macdonald says of learning of her late twin brother: "When I found out about my twin many years later, the news was surprising. But not *so* surprising. I'd always felt a part of me was missing; an old, simple absence. Could my obsession with birds, with falconry in particular, have been born of that first loss? Was that ghostly kestrel a grasped-at apprehension of my twin, its carefully drawn jesses a way of holding tight to something I didn't know I'd lost, but knew had gone? I suppose it is possible."[13] Macdonald's early fascination with birds developed into her becoming a falconer, one who trains birds of prey that fly freely yet are tethered—literally during training, and metaphorically when trained—to the human with whom they are deeply connected.

Whether they are monozygotic twins or dizygotic twins, or brothers or sisters born or adopted months or years apart, the great majority of siblings grow up together. And because they grow up together, they develop a codependent relationship that more often than not is beneficial and constructive. Most grow up with the same family mores, and in the same culture.

When they develop into young adults, the concurrent life that siblings lived together may come to an end. They may leave their family of origin to live an adult life that fulfills their personal, social, and professional dreams. But while the tightly knit years of growing up together are gone, the collective history of siblings is embedded in their neuronal and kinesthetic memories. The genetic and experiential elements that twin and twin, sister and brother, brother and brother, or sister and sister share will remain pivotal throughout their lives. When one of them dies, the loss seals off access to the world they shared with each other growing up—to memories of being kids together and to being part of a family for all those years, and for all those shared experiences. Losing a sibling, then, is not like losing a piece of yourself. It *is* losing a piece of yourself.

ELEVEN

Life Partners

Grief was not like anything I had imagined.
—JONATHAN SANTLOFFER

The grief after losing a spouse or significant other can be crushing. When someone's "person" dies, every part of a life lived together, the many treasured and unique things that formed their all-encompassing shared life—countless shared experiences and memories, a deep intimacy, laughter, wrangles and prickly times, planning the future, joking, laughing, kissing—screeches to a halt. No longer is there physical and emotional intimacy, part of which was their erotic relationship, and for many, the drive to begin a family. The physical touch from one's beloved—part of a couple's intimacy, the loving bond that animates the physical and emotional exchanges between partners—is no more, and will be missed, always. The special ways a couple loved each other, at different times over the course of their relationship, are gone. The death of a wife or a husband has been described as among the most stressful of life events.[1]

Julian Barnes captures the depth and complexity of this relationship when he questions what exactly he is missing after his wife dies. He wonders whether he's "missing her, or missing the life we had together, or missing what it was in her that made me more myself, or missing simple companionship, or (not so simple) love, or all or any overlapping bits of each? You ask yourself: what happiness is there in just the memory of happiness? And how in any case might that work, given that happiness has only ever consisted of something shared? Solitary happiness—it sounds like a contradiction in terms, an implausible contraption that will never get off the ground."[2] In grappling with what his wife's absence means, Barnes tries to understand the effect her death has had on his physical and psychological experience of life.

Spouses

The time between Barnes's wife's diagnosis and her death was brief, just thirty-seven days. For some spouses or significant others, however, the death of the other comes suddenly in the form of an accident or a medical emergency that fells the body in an instant.

A colleague of mine suffered the anguish of her husband's sudden death, when they were together in their suburban home. Weeks later, she told me that she and her husband had woken up early, even before the sun had fully risen, for their morning run.

They both remarked how vivid the pre-dawn sky looked, streaked with orange and red. It was after those words that her husband fell, dropping onto the wooden floor in the foyer. Sometime later, at the hospital, she remembered the mariner's rhyme that forecasts a storm: "Red sky at morning, sailors take warning."[3] Words of a verse meant to signal only caution, not prevention. Nothing can prevent a storm. My colleague's husband's sudden and unexpected death, from a massive heart attack, came without any warning; there'd been no symptoms. It didn't make sense, she said.

Another husband, the father of a close friend, died quickly of a sudden stroke. The tragedy happened late one evening. My friend's father was an opera buff, and her stepmother was a baseball enthusiast. The couple had returned to their apartment after the last opera performance of the fall season: Verdi's *Rigoletto*. Years before, soon after they married, they had struck a bargain: she would accompany him to the opera performances he loved, and he would accompany her to her baseball games. The incongruity amused the family because of the juxtaposition not only of their gender-atypical preferences, but also of their physical appearance: he was six-three and heavy, and she was five-three and slender.

The walk from the opera house had been cold, and the two made hot cocoa to warm themselves. They sat in the kitchen, drinking it, when suddenly he fell forward, his head coming to rest on the kitchen table. His wife called 911 immediately. But it didn't matter: the cerebral hemorrhage was fatal.

Anguish from emptiness, from loneliness, can reach into the deepest parts of the self. There are widowed ones whose eyes continue to fill with tears whenever their deceased spouse comes up in a conversation. Their time of mourning remains fresh and without words—at least not yet. The brain and the body strain—yearn—to regain the grasp of a hand, to feel a kiss, and to be held in an embrace. The longing is for any familiar physical contact that can make the brain respond as it had in the past to feeling the loved one's touch. It is the aching and the longing for the lost one's kinesthetic expression of affection. The physical and psychological attachment each shared with the other turns into a loneliness suffered by the surviving partner.

In a long-term relationship that becomes a lovingly solid one, the activities and understanding the two people share develop beyond the initial passion of sexual intimacy. Physical knowledge and psychological familiarity lead to a deeper intimacy: Whatever the sexual orientation of partners—lesbian, gay, bisexual, or straight—a long-term relationship can achieve such closeness that when one partner thinks a thought, the other will verbalize it. And when it happens, both will laugh, enjoying the playful idea that thoughts could be broadcast from one to the other. From a psychological perspective, this is similar to priming: something, such as a song or a book title, is recognized and triggers a particular response, which in this case is familiar to both people. Sometimes just spending time together and coming to know one another so well creates situations in which both members of the couple have identical responses to things.

Attachments and Grief

Much of the experience of grief is shaped by the attachment history of the bereaved. If the survivor's early attachments were insecure, then grief will be harder to manage and endure. But even for those survivors, if later relationships offered the emotional space needed for them to become more secure, they will likely have gained some psychological strength to help them tolerate the separation of death. Other circumstances that influence the post-burial grief of surviving partners are the personal and social resources available to them.

The form of grief that a bereaved spouse experiences will also depend on the nature of their relationship with the deceased. The length of time a couple was together, the quality of their day-to-day life, and how they dealt with their emotions are all relevant factors. How accessible to each of them were emotions that influenced whether interactions were generally loving, prickly, or angry? Were the two often polite, but distant and unable to express what they were feeling? Each partner has an effect on the other's quality of life as well as on their own psychological functioning. In couples whose quality of life encouraged personal growth and interdependence in their relationship, research shows that this quality of life will continue after a partner's death.[4]

Widows, Widowers, and Others

Widowhood has no exit. Yearning, sometimes searching for the lost one, thinking the loved one will return: they know this is irrational. Death and everything that follows in new widowhood seems irrational. Words, thoughts, don't work. The brain struggles with what "forever" means. How can the heart feel "never"? "I lost my wife"—the widower's words. "I lost my husband"— the widow's. Four words that for each fit heartbreak into one sentence.

But for those who are unmarried and have not been together for many years, there are no identifying words to proclaim the profound loss of a beloved partner. Megan Devine, who witnessed her partner's death by drowning, writes, "With his abilities and experience, there was no reason he should have drowned. It was random, unexpected, and it tore my world apart." In a *Washington Post* article, Devine described the nebulousness of being outside the norm, the "erasure" that happens when one loses an unmarried partner.[5]

Identities change when a spouse dies. A wife's identity changes from wife to "widow," while a husband's identity changes to "widower." Unless a spouse remarries, these terms mark a new sense of self, a new identity. It's different when parents lose a child: parental grief is immense, as with the widowed, but there is no analogous term to identify bereaved parents or, for that matter, a bereaved unmarried partner.

Those whose lives were organized around a partner's life are often more deeply affected by their death and will experience more psychological distress.[6] And for some widows and widowers, the psychological stress can be expressed physically when the grief isn't fully experienced. When sorrow has been cast into an indirect form of grief (such as forbidden or delayed), that grief will likely find a way to be expressed through the body, which always remembers what happened.[7] Any number of a widower's or a widow's functions and organs may become compromised, not least the heart itself. In extreme cases, it's possible that the death of a significant other may be so devastating that it hastens the survivor's own death.

But there is Eros, both the life force and the love force (described by Freud and the ancients), and there is the survival instinct, which pushes most who are bereaved to recover from the shock of the death and its aftermath, whether the death was expected or not. When grief can run its course, it will eventually calm. And the will to live a life that has changed will take over.

Last Words

Joan Didion's husband died suddenly, reading a book as she was tossing their dinner salad. As Didion describes that one moment: "John was talking, then he wasn't." She looked up from the salad when he stopped talking. She writes, "Life changes in the instant. You sit down to dinner and life as you know it ends."[8] Later, when she tries to remember what his last words were, she's not sure. Were the words about the World War II book he was reading, or were they about the scotch he was drinking? When her husband's words stop, Didion sees him "slumped motionless." She calls the number they have at the ready for others in the apartment building if needed, for the New York-Presbyterian ambulance. Didion watches the hospital paramedics arrive, but is not sure how many there are. She watches as they move through their choreography of procedures that turn her living room into a medical emergency station. The paramedics try to restore a heartbeat with defibrillating paddles. And with injections of something she's not sure of. They can't. Nothing works. Didion's husband does not respond. She follows the ambulance to the hospital, where a man in the hospital driveway, the only person not in scrubs, asks, "Is this the wife?"[9] Hearing him tell her that he is her social worker, she knows her husband has died. Didion has moved into the world of widows.

When she writes about the trauma of her husband's abrupt death, Didion's prose cuts with precision. "My husband, John Gregory Dunne, appeared to (or did) experience, at the table where he and I had just sat down to dinner in the living room of our apartment in New York, a sudden massive coronary event that caused his death." After her husband died, it was only during that first night that Didion felt she had to be alone, "so that," she wrote, "he could come back." Over the course of the year following her husband's death, Didion wrote *The Year of Magical Thinking*, a memoir of grief.[10]

It was dusk when a friend of my husband's slipped in their driveway on ice, black and unseen in the twilight. She wasn't able to get up. The medical report after the fall was unexpected. MRI scans of her spine taken after she was brought to the hospital showed that cancer had penetrated her spinal cord. Her husband and other family members begged her to undergo chemotherapy, but she refused. She told them, gently but firmly, that the quality of her life, given the grim prognosis, would not be worth it. She had a favorite way of looking at actions from a cost-benefit perspective. For her, chemotherapy treatment would result in a negative balance, and she wanted to have some control over her death. Her husband—loving her, talking with her, knowing and understanding her—finally agreed.

His wife began writing goodbye notes to her loved ones in the hospital, and finished them at home before she was taken into hospice care. Six weeks after her fall, she died. Her husband, a resolute soul, told us he was shattered. They had been married for forty-five years.

Antoine Leiris, a journalist in Paris and another husband who was shattered by the death of his wife, writes of his unwillingness to give up the present tense in which he was living just before his wife was killed. Leiris's heart-wrenching memoir reflects his shifting psychological states beginning on the day of his wife's death. At first, he hears of a terrorist massacre at the concert his wife was attending. When he learns that Hélène was among those murdered, he tries desperately to stay in the present by using the present tense: "I had to write quickly. I am the one who loves Hélène, not the one who loved her. Before death drops the curtain down on the person I was, without a final encore, I am still this naïve fool prevented from falling by hope. Who knows what I'll become tomorrow, when my grief has let me fall?"[11]

Leiris's hold on the present, his conscious denial of his wife's death, soon changes. He continues, "Like a brief, fleeting love, an all-consuming passion, it is just passing by. A vivid reflection of the love that shone so brightly. It is beautiful, intense, I hold it tight against me. But I know that it has almost left me already. In search of another lover to torment, it goes on its way, abandoning me to its sad traveling companion. Mourning."[12] Leiris begins his memoir almost immediately after he learns that Hélène may be one of the victims of the massacre. His arresting prose describes the frenzy of his emotional attempts to stay in the present. But the defense against reality quickly diminishes, succumbing to the inevitable slide into grief, new and stunned by shock.

Grief-Stricken Royals

The historical examples of the British royals King Edward I and Queen Victoria, described earlier, illustrate how each dealt with the sorrows of their widowhood. In the nineteenth century, Queen Victoria's grief after her husband Prince Albert died never ended. For three years she refused to go out in public. The "Widow of Windsor," as she became known, unable to let go of her grief, refused to change anything in Prince Albert's rooms except his everyday things: clothes laid out for him, shaving gear, and table settings. Queen Victoria dressed in black for the remainder of her life, and never married again. She survived her husband by forty-one years, and died in 1901.[13]

In contrast to the relentless, complicated form of Queen Victoria's grief, King Edward I expressed his grief by erecting crosses to commemorate his wife, Queen Eleanor, at each place where the funeral cortège stopped overnight. King Edward remarried in 1297, seven years after the death of Queen Eleanor, and lived for ten more years.

A Widow and Motherless

During our first meeting, a patient described herself as pretty new to widowhood. How Julianna's grief tumbled out—as I look back—was a classic example of the surprises that grief can have in store for the survivor. Helen Macdonald speaks to what can lurk, muddled and hidden, beneath sorrow: "The archaeology of grief is not ordered. It is more like earth under a spade, turning up things you had forgotten. Surprising things come to light: not simply memories, but states of mind, emotions, older ways of seeing the world."[14] The overwhelming and dimensionless nature of grief can bring up all kinds of emotional material, some of it long forgotten, before diminishing and then changing.

In sessions Julianna would often sit and just cry. At times, too, she would describe the agony of missing her husband and how at home, becoming acutely aware of his absence, her heart would ache. But at other times she expressed grief in an unusual way: she would become intensely angry—almost fly into a rage. The intensity of this anger when it erupted made me think that this widow's suffering included something beyond her grief in the present.

In fact, the rapidity of Julianna's emotional flight away from her grief, and into anger, felt to me like being in a bate. In falconry when an accipiter bird, a sparrow hawk or a goshawk, wants to escape its trainer, it will fly into a bate, which T. H.

White defines as a "headlong dive of rage and terror, by which a leashed hawk leaps from the fist in a wild bid for freedom, and hangs upside down by his jesses."[15] To me, it seemed as if Julianna wanted to fly from grief, down into something deeper than the present, something terrifying. She would cry and talk, and sometimes she paced while she cried. Her words and tears poured out. We understood that her widow's sorrow was being verbalized and cried out. But there was something else igniting her angry outbursts, which would be followed by long silences spent staring out the window.

For Julianna, it turned out that what was hidden from her conscious self was the death of her mother, which had produced a grief that had subtly stalked her consciousness for years.

Discoveries like Julianna's can happen in therapy, when a person's defenses become porous enough to allow access to forbidden memories. Julianna's feelings were monumental, and at first she couldn't find words for them. Finally her anger, desolation, and hopelessness surfaced often enough that she could recognize and face her feared and forbidden grief. She began to talk about her mother and the details of her mother's death.

Julianna's mother did not suffer a heart condition and die from it, as Julianna's husband had. Her mother had committed suicide. Julianna was at her best friend's house—it was mid-morning during a school vacation—when her younger sister, Genevieve, called. Her sister sobbed the words, "Where have you been? I've been trying to find you. Mummy's dead. Daddy told me to find you and tell you."

To remember and listen to herself tell me the full story of her mother's death was, she told me, an epiphany. She was able to put together information and details "that were scattered until now." She knew her mother had health problems, but her parents never discussed what they were. She wondered if any of those problems had to do with mental health, but there was never a time before her father died years later that he could talk about it. Whenever his wife's death came up, he would tear up and tell his daughter he couldn't bear to talk about her mother. Her father's grief was chronic; Julianna's was forbidden, and became inaccessible.

Julianna once thanked me for meeting with her in mid-morning. She had always dreaded that time of day. The middle of the morning during the week inevitably made her feel unpleasant. And after her husband died, it became the worst time of day. She told me her grieving was like what she remembered in the poem, "Ode on Melancholy," by John Keats: that sorrow needed a partner.[16] Julianna said I was her partner.

It was clear why mid-mornings were painful for this widow. In grieving her husband, who had died at ten o'clock on a Thursday morning, her forbidden and unaccessed grief after her mother's death surfaced—and this grief, too, was associated with morning. Everything began to spill out, as Julianna began to understand her own feelings. There were multiple parts to her anger: that her grief was forbidden; that her sister had blurted out the news; and that her father had refused to talk about her mother's death. The enormity of the loss of her husband tore at her, but it also opened a way to reach the grief of her mother's death. These losses and the accompanying anger engulfed her—for a time—like a tidal wave. She grieved the husband she loved immeasurably, and eventually she was able to move beyond her anger at her mother for committing suicide, remember how much she loved her, and grieve her loss by reappropriating that grief and making it her own after all those years.

Julianna's experience was an amalgam of different kinds of grief, mixed with anger and feelings of abandonment. I watched, helped, sometimes felt helpless, and marveled at Julianna's courage to confront so many powerful feelings. Her claim to this anger, grief, and love gave her the strength to resume living her life—a different life that was made possible when an active grief unearthed a forbidden one.

She acknowledged the damage that her mother's death had done to her, as well as how hurtful it had been to have her father put up a wall because he could not deal with his wife's death. What for so long seemed like her father's tearful refusal to talk, she realized, was actually an attempt to avoid his own emotions aroused by his wife's suicide. Unable to talk about his conflicting feelings—such as anger, trauma, sadness, guilt, and confusion— he was never able to face those emotions and was never able to grieve her death.

At the end of therapy, Julianna said the story of her life had changed: her grief, once forbidden, had become accessible. Grieving her husband was what led to grieving her mother's death.

Memories of her husband and her mother were mixed with sadness, and with joy. She had claimed a grief that had lain dormant and disturbing for years, and now her life slowly was becoming enriched.

Identifying with Those Lost

As mentioned earlier, grief that is blocked from emotional expression may be experienced in the body as physical symptoms similar to those experienced by a loved one, especially when the relationship with the loved one was very close. What Americans term "facsimile illness," Darian Leader, a psychoanalyst, calls "homeopathy"—in the British sense of the term—to describe a way the bereaved may identify with a deceased loved one. Identification in this way with the lost loved one can take different forms: by contracting a similar illness, for example, or by "taking on aspects of their behavior, [or] mannerisms." And it doesn't stop there: survivors can even take on their lost loved one's "ways of looking at the world."[17] Leader writes, "A woman mourning her husband not long after his death noted how, when confronted with a problem, 'I deliberately looked at this in a way that my husband might have done had he been alive.' . . . She had turned it into an even more successful enterprise, emulating not only her husband's interests but his ways and methods of handling business matters." As some behaviors in the bereaved can imitate those of a lost partner, physical symptoms in the bereaved may also mimic those endured by the lost partner.[18]

Unexpected or Anticipated Heartache

How the death of a loved one happens is every survivor's story, recalled as memories and related as details that swirl into conversations and speeches at the after-events. Unexpected deaths call for different stories than those following a terminal illness.

An unexpected death can be caused by a natural death blow, such as a heart attack or a stroke. It can be self-inflicted, as in a suicide, often unexpected even by those close to the deceased. Or it can be caused by some unanticipated external event: an accident such as a car or plane crash, a homicide, or a terrorist killing.

At a concert in the City of Light, Antoine Leiris's wife and more than ninety others were murdered by terrorists. Some years earlier, a lone shooter in the United States massacred twenty children between age six and seven, and six adult staff members, at Sandy Hook Elementary School in Newtown, Connecticut.

In Paris, Hélène's death left Antoine Leiris spouseless, and left his toddler son motherless. In Connecticut, the deaths of children and adults left spouses, siblings, and parents bereft with sorrow. The horror and the shock of these mass murders captured the attention of the world. Leiris, a widowed survivor, expressed the totality of his early anguish with words that squeeze the reader's heart in *You Will Not Have My Hate*, his memoir.[19]

But if a death is predicted, and therefore expected, what is the heartache that surrounds the spouse who feels anticipatory grief living with their partner until they die? When a diagnosis is given that demarcates the probable time left before a beloved one dies, a space opens for death to hover before making its claim, and for hope, which springs as eternal as we are human, to crowd in— hope in the wish to override medical evidence and offset the expected. Hope for more time, more life with the loved one.

C. S. Lewis and his wife both knew she had terminal cancer. Lewis writes of the experiences of his mother, his father, and then his wife, all of whom died of cancer.[20] His wife, whom he calls "H" in his writings, has cancer that remits briefly. During this brief and loving time, Lewis and H laugh and talk and live their lives with as much happiness and wonder as possible given the ominous possibility that the disease might return.

Afterward the cancer swoops back to regain a firmer and final clutch on her life. Lewis groans with grief upon H's death: "And that, just that, is what I cry out for, with mad, midnight endearments and entreaties spoken into the empty air."[21] He writes that "bereavement is a universal and integral part of our experience of love," adding that he is concerned he will forget her. Yet he goes on to write, "the remarkable thing is that since I stopped bothering about it, she seems to meet me everywhere." C. S. Lewis, in his grief, expresses its torture, its relentless attack on the mind and the body. He comes to a place where grief's diminishment harkens its change, and his.[22]

Julian Barnes, whose wife died just over a month after her diagnosis, also lived with his spouse for a time knowing she would succumb to cancer. Although the times of their wives' pre-death periods are different, the anguished grief of both Lewis and Barnes is similarly described. Grief is the price of love. C. S.

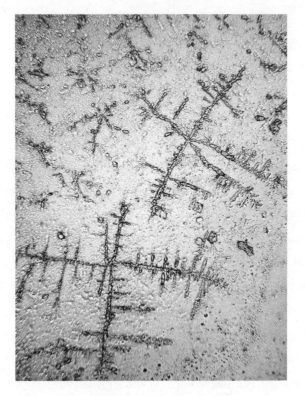

After goodbye. From *The Topography of Tears*, © 2017 Rose-Lynn Fisher,
published by Bellevue Literary Press.

Lewis writes of grief as a cost the bereaved pay for their love.
Barnes notes that although love may bestow a sense of invinci-
bility on lovers, "there is always the sudden spear-thrust to the
neck. Because every love story is a potential grief story."[23]

What is grief like after the final goodbye? How much time
will it take for grief to loosen its grip after a partner dies?

At first, C. S. Lewis thinks sorrow is a state, and that he "could
describe a *state;* make a map of sorrow. Sorrow, however, turns

out to be not a state but a process. It needs not a map but a history."[24] In continuing this discourse, he realizes that at some arbitrary point his history of grief needs to stop because, he writes, "There is always something new to be chronicled every day. Grief is like a long valley, a winding valley where any bend may reveal a totally new landscape. As I've already noted, not every bend does."[25] Lewis, like Julianna, Helen Macdonald, and others, is struck by surprises during the time he is grieving: "Sometimes the surprise is the opposite one; you are presented with exactly the same sort of country you thought you had left behind miles ago. That is when you wonder whether the valley isn't a circular trench. But it isn't. There are partial recurrences, but the sequence doesn't repeat."[26]

Julian Barnes writes of his experience of time as he grieves his wife's death: "The grief-struck themselves can hardly tell [time is passing], since time is now so much less measurable than it used to be. Four years on, some say to me, 'You look happier.'" He comments, too, on the language of grief, on the words that for him are part of a different vocabulary. After four years, he is struck by the fact that the grammar he uses, "like everything else, has begun to shift: she exists not really in the present, not wholly in the past, but in some intermediate tense, the past-present."[27]

How soon grief will change cannot be predicted, and the pace of its changing cannot be hurried. Jane Brody, a *New York Times* columnist, writes that the time it takes to recover varies for everyone who is grief-stricken. She takes issue with previous work saying that 60 percent of people recover within a year after losing a spouse. Newer data, Brody writes, suggest that it can take "two to three years or even longer for some to recover from bereavement."[28] The length of time Brody cites is comparable to the time Julian Barnes says it took for him to enter his front hall without crying: going on four years. But there is no timetable for the griever's grief. Only the griever will know when sorrow changes. Yet there are signs that the change is on its way. One reliable sign of grief diminishing is the ability of the bereaved to accept condolences gratefully. Another is being able to think of the deceased without the "wrenching quality it previously had."[29]

Results from research studies are continuously improving our understanding of the psychological and physiological underpinnings of grief, including how it changes the brain. But one person—just one individual—is not a statistic. No grief-stricken wife, husband, or partner can be shoehorned into the statistical figures that are displayed in a table. Research data give us collective information about averages and outliers, not the story of a particular spouse or significant other.

Life as it was, before the death of a spouse or partner, is over. As sorrow takes hold, then changes, the bereaved survivor creates a new life, one that has no blueprints, statistics, or timetable.

A Bittersweet Alchemy

I loved this world very much,
The trees, the wind, the mountains, the air.
So every time you are out of doors, close your eyes,
Take a good, deep breath, let yourself really smell the air,
And know that I am with you then also.
—LETTER FROM A GRANDMOTHER TO AN UNBORN
GRANDCHILD

Long ago, alchemists famously pursued the wild and tantalizing dream of changing lead into gold. What those proto-chemists didn't know was that to achieve this "fabled transmutation" would require moving protons, which isn't possible by any chemical means (and is impractical even with today's supercolliders).[1] The reasons their experiments failed wouldn't be known until centuries later. But despite the futility of their efforts, there was an important outcome: many of the experimental methods developed by sixteenth- and seventeenth-century alchemists laid the foundation for modern-day chemistry.

The heartache after a death is not anything we can see or hold in our hands: it lacks substance. Yet this emotion, experienced by those whose loved ones have died, has sometimes been compared to lead. The grief-stricken describe feeling leaden and heavy, and say that their world has gone gray. No color, just a washed-out, leaden landscape that they face day after day, week after week. Grief can be as intense and long-lasting as it is because, however the bereaved experience it (as mild, moderate, or severe), and whatever form it takes, it perfuses the human self. The effects are all-encompassing—it changes the brain, hurts the heart, and makes the body suffer.

And the quality of grief? Not gentle like mercy, that quality Portia so beautifully describes: "The quality of mercy is not strain'd, It droppeth as the gentle rain from heaven."[2] And not sluggish, or inert. Grief is dynamic. It can rumble and then be calm for a time, only to swirl up unexpectedly, even furtively. Emerging in one form, it can morph into another. Whichever of the many forms grief may take, and whatever name belongs to the form—chronic, ambiguous, resilient, or delayed—it affects the whole self. Dismal feelings, sad thoughts, sighing, crying—internal phenomena, externalized—all circle back to the loss of the deceased. As days, weeks, and months clatter on, it's as though grief refuses to change. Yet it can—the capacity for change is there in the bereaved.

But there is a condition. For grief to change, it needs the freedom to flow. And when it is permitted, even invited—through knowledge—grief as lead can be changed into joy as gold.

The jazz singer Diana Krall, interviewed after the release of her album *Turn Up the Quiet*, talks about what the album represents to her. She made the work as a tribute to lost mentors and loved ones, and as a remembrance of them. Krall had "lost a stream of close relatives and mentors, and . . . the album is a reflection of her progress in dealing with grief."[3] She describes coming to the realization that in enduring grief, finding joy is possible: "It's not about a period of time or a demographic. It's about finding romance in everything, in beauty or in things that are sad."[4] Her songs tell of suffering and braving grief, and then finding joy.

But first there is grief. Lead before gold.

In the real world, lead cannot be changed into gold. But such transmutation can occur in fairy tales, stories that have emerged and been retold countless times because of their representative truth. Consider "Rumpelstiltskin," a story thought by some researchers to reach back four thousand years.[5]

> Once upon a time, as the story begins, a miller was overheard saying that his daughter could spin straw into gold. Word of this reached the king, who summoned the miller and commanded him to prove it, locking the miller's daughter in a room for the night with a pile of straw and a spinning wheel. The poor young woman was terrified, for she could spin wool into yarn, but not straw into gold. Certain of failure, she was amazed when a goblin suddenly appeared before her. He offered to spin the straw into gold for a price, and she agreed,

paying him with her necklace. The goblin kept his end of the bargain, spinning the pile of straw into gold overnight.

The next day the king unlocked the door and was delighted upon seeing the gold. He arranged for a larger pile of straw to be placed in the room and asked the girl to perform the feat again, locking her in. That night, the goblin appeared a second time to the frightened young woman, saying that in return for the ring she wore, he would spin the large pile of straw into gold. She felt she had no choice but to agree.

On the third day, the king took her to a much bigger room in which straw had been piled to the ceiling. He promised to marry her if she succeeded in spinning all of the straw into gold, and to lop off her father's head if she failed.

That night the young woman, tearful and trembling with despair, waited for hours before the goblin finally appeared. He immediately demanded payment for spinning straw into gold for a third time, but the young woman had no more jewelry to offer him. So the goblin proposed something else. He agreed to spin the straw into gold if the young woman would promise to give him her first-born child.

What could she do? She agreed, sobbing, and the goblin performed his magic a third time.

The next morning the king was elated. He married the miller's daughter, and everything went well until the woman, now the queen, gave birth. The goblin appeared in the nursery one morning and demanded that the queen honor her end of the bargain, and give him her baby. Stricken, she pleaded with him: how could she possibly give up her first-born child?

The goblin relented, but only a little. He gave her one way, only one way, to get out of the terrible bargain: she must guess his name. Panicked, the queen sent her most faithful servant on a mission to discover the goblin's name. Trekking deep into a thick forest far from the castle, the servant came upon the little goblin dancing around his campfire, chuckling with glee and singing, "Rumpelstiltskin is my name!" The goblin repeated his name over and over, exulting in the certainty that the queen would never guess it. The servant hastened back to the castle.

The next morning, Rumpelstiltskin appeared before the queen and demanded the baby. But first, what was his name? After two tries pretending she didn't know, the queen asked the little man if, perhaps, his name could possibly be Rumpelstiltskin?

She had named him. It was no guess—knowing his name, she had become empowered. Rumpelstiltskin fled, defeated. Naming the goblin rendered him powerless to tear her family apart.

The Elf. "Rumpelstiltskin is my name!" From *Anne Anderson's Old, Old Fairy Tales*, © 1935 Anne Anderson, published by Whitman Publishing Company.

"Rumpelstiltskin" dramatizes how powerful knowledge can be. This happens in life. To learn and understand is to acquire knowledge, to make what is unknown known. The fairy tale that tells of straw being spun into gold is a fantasy. But what is not fantasy in the story is the power of naming. To name what is unnamed can liberate the self from the mystery of what is unknown or forbidden. To name the unnamed is to gain the power to change the course of one's life.

A Slow Metamorphosis

The internal effects of grief aren't visible. Neuronal connections change as places, people, and things become experienced in new ways without the loved one, yet often there are no words. Language falls short in the beginning. The bereaved is hard-pressed to convey how heavy the heart feels, or how the body aches. The brain's somatosensory cortex, where the loved one's touch was registered, no longer responds the way it had. Something is missing: the loved one.

The bereaved knows this cognitively, understands it, but it's as if at times what is missing refuses to be named. "Nothing makes sense anymore," some say. It is not only that the bereaved need "to do grief-work," as others in addition to Freud have written, but that grief (intangibly) needs to do its work on the bereaved. Both grief and the bereaved need to change.

When grief has the freedom to be expressed—without avoidance or denial—it takes over and pursues the path it wants to take. No map, no predictability, no stages: grief is its own entity, and has its own compass. Sorrow keeps spreading. Its presence is leaden, a gray mantle on the bereaved.

Yet without any foreshadowing, without any announcement, the tight space that death forced the mourner to occupy loosens. There is an opening—slowly, without any sound or awareness—into another place. At first, the opening is narrow, but gradually it widens as wonderful, comic, loving memories of the lost loved one emerge. This is how grief, allowed to go where it will, changes. Different behaviors without the loved one are at first repeated with conscious and active effort, but gradually, new experiences and skills become familiar.

Henry Marsh, a British neurosurgeon, describes how this transformation takes place: "When we learn a new skill the brain has to work hard—it is a consciously directed process requiring frequent repetition and the expenditure of energy. But once it is learnt, the skill—the motor and sensory coordination of muscles by the brain—becomes unconscious, fast and efficient." He continues, "To learn is to restructure your brain."[6] As new behaviors are learned, and the brain changes, so is the loved one remembered.

As life without the deceased is lived through grief's vicissitudes, a shift begins. Tears diminish. Smiles, tentative at first, but real, widen and last longer. Attention is caught by something interesting. A newspaper, an article, a book is read with surprising clarity, rather than with lapses in concentration. This is grief ebbing, and without conscious intent. As Julian Barnes puts it, "All that has happened is that grief has moved elsewhere, shifted its interest. We did not make the clouds come in the first place, and have no power to disperse them. All that has happened is that from somewhere—or nowhere—an unexpected breeze has sprung up, and we are in movement again."[7]

Eventually there is the realization that life is being lived without the loved one. The bereaved come to understand that they won't forget the loved one if they feel pleasure and joy again, that it's all right to be happy. This is the sense of grief moving away, subtle but sure. Despite occasional bursts of intense grief, more and more positive times and experiences are remembered and lived.

Yes, though sadness is the nature of grief, that nature can change into a kind of nobility. And grief as despair and sadness can be converted into joy when memories of the loved one, bittersweet but joyful, begin to bubble up. With grief ennobled, pleasure and joy return, felt differently than before but just as truly. The pain of grief, felt and endured, allows that space of deep sorrow to be reappropriated by other emotions and filled with the memories of the one whose life brought so much happiness.

In "Finding Joy in My Father's Death," Ann Patchett writes of her experience during the three years her father lived with a degenerative neurological disease. She runs into a friend at a store, telling her friend that her father is dying. "He's still alive," she explains, saying, "I've decided to wait and feel terrible once he's dead." Her friend responds, "Or not," and hugs Patchett. After her father dies, Patchett realizes how wrong she'd been. Because it wasn't just when she visited her father that she felt sad; she was sad all the time.[8]

She describes going beyond her own expected response when

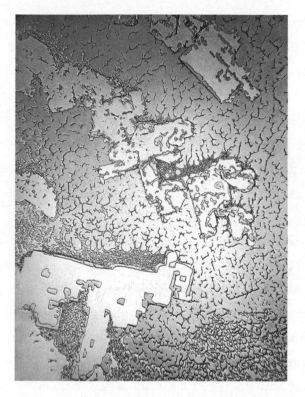

I remember you. From *The Topography of Tears*, © 2017 Rose-Lynn Fisher, published by Bellevue Literary Press.

she expressed that death and joy could be experienced together. Having endured what seemed like anticipatory grief for several years, she wrote that there was "joy in the place" her father left after he died.[9]

Emblematic of the deep love for their lost one, the bereaved reach a place where, paradoxical as it may seem, they are enriched by grief as they realize that death and joy can exist together. Many who have allowed grief to run its course are surprised by the shift out of the sorrow they've been living in. When grief is ennobled, past joys can emerge in the memory and heart of the bereaved.

The patient of mine who had experienced a complicated and prolonged grief after her husband died, and who eventually realized that an unopened chamber of grief had been created by her mother's suicide years before, smiled in a session one morning, saying her shoulders weren't tense and her chest didn't feel heavy.

Another patient, whose losses included not only her father several years before, but recently her mother, came in one afternoon with a lighter step. She smiled a real smile. She stated how important it was that she could acknowledge how full of conflict the relationship between her and her mother had been. Anger and sadness, she said, had somehow gotten stuck together: anger had fused with grief, getting in the way of her ability to mourn her mother. In facing her anger, memories of wonderful times with her mother surfaced. She described a favorite memory: shopping with her mother for her first prom gown. When she had tried on a gown that made her feel like a princess, her mother had smiled with tears in her eyes. That day, with its long-forgotten feelings of admiration and motherly love, was reappropriated. She got back that day, and others like it, to remember with pleasure.

In her intake interview, an articulate woman said she'd been pregnant, but that was a long time ago. As her therapy sessions unfolded, she became comfortable describing the many successes of her professional life and the history of her family. The pregnancy continued to be referred to only as a pregnancy in the infrequent times it came up. Then one day, in talking about grief in general, something different happened in the discourse. The word pregnancy changed to the word baby. She said she'd had a baby, a girl, who had died ten minutes after she was born. To be able to say that she'd had a baby who died was a sea change, much like Ariel in Shakespeare's *The Tempest* who sings to Ferdinand whose father had drowned, "Nothing of him that doth fade, But doth suffer a sea change into something rich and strange."[10] The power of using the word "baby" was key in connecting what really happened to the part of herself that had needed to keep that loss at a distance, but now was ready to grieve.

Change was evident, too, in the poignant expression of bittersweet joy articulated by the woman whose archer-son was killed in a vehicle crash. Her heartfelt words expressed what she could allow herself finally to feel after months of traumatic and complicated grief combined with survivor's guilt: an ennobled grief, felt, explored, and cried over. She described how even though she hadn't wanted to move away from the tragedy, time and her work in therapy had pushed her ahead. She pressed her hand to her heart, patting it, remembering how courageous and happy her son always was, and said that he now lived in a cherished place in her heart. Then she smiled.

Transformations for the Fatherless

A patient recalled a loving memory of her father that was precious to her. She said she had reached—finally—an understanding of how full her father's life had been, and why he had been so seldom available to her. She described her father's enthusiasm for life, and his laughter, and how that translated into his delight in loving her. She remembered how, after he came back from a trip (he traveled internationally), she would run to him, and he would swoop her up into his arms as he laughed and hugged her.

In Helen Macdonald's *H Is for Hawk*, months after her father's death, she stumbles upon a key taped to a cardboard that has been pressed between two of her father's notebooks on a bookshelf. As she sees it, memories tumble out from long ago when her father took her to visit England's Stonehenge. In the mental image she sees of herself, she is a very little girl holding her father's hand, looking up and asking him if the stones with no walls behind them were an entrance to a house.

"'Is it a house, Daddy?' I asked him. 'No one knows,' he said. 'It's very, very old.' I held the cardboard and felt its scissor-cut edge. And for the first time I understood the shape of my grief. I could feel exactly how big it was. It was the strangest feeling, like holding something the size of a mountain in my arms. *You have to be patient*, he had said. If you want to see something very much, you just have to be patient and wait. There was no patience in my waiting, but time had passed all the same, and worked its careful magic. And now, holding the card in my hands and feeling its edges, all the grief had turned into something different. It was simply love. I tucked the card back into the bookshelf. 'Love you too, Dad,' I whispered."[11]

Daughters and sons who grieve—deeply, expansively, angrily,

lovingly—however long it takes, allow their mourning to leave center stage. Joy begins to glide into the limelight, propelled by the knowledge of having loved a father now deceased, and of his having loved you.

Expressions of Grief

It takes—how long? For some, a very long time. There is no timetable, only patience and the surety that grief will let go of the griever.

But until that time is reached, what about the vacated space? Grief's space, empty and echoing, vacated by the loved one: how to fill it? Works of art created throughout history in architecture, music, painting; memoirs; snippets of prose or poetry on three-by-five cards placed inside a book being read, or in a notebook with pockets—anything created by the griever that speaks to the lost loved one who will forever be missing, can work in service of the sadness. Whatever can be created to fill the space vacated by the loved one will be essential and meaningful.

Even the simple act of speaking, using words to name unnamed feelings and thoughts that swirl through the bereaved, helps. On the power of speech, Darian Leader writes, "Drugs can alleviate surface pain but they cannot affect personal, unconscious truth, which can only emerge through speaking."[12] Leader also stresses how important art is in expressing grief: "Works of art, after all, share something very simple: they have been *made*, and made usually out of an experience of loss or catastrophe. Our very exposure to this process can encourage us, in turn, to create, from keeping a journal to writing fiction or poetry or taking brush to canvas. Or simply to speak and think."[13]

Backward and Forward

It isn't steady and straight, this thing called grief; it goes every which way unpredictably. There are innumerable metaphors—a surplus of them—that can illuminate its fickle nature. Take one: powdery snow in high drifts swirls into shapes driven by gusts of wind. Or another: a pendulum swings again after the clock is wound and resumes its steady "tick tock," time and life swinging together, only to wind down again.

Julian Barnes is aware of how grief, seemingly gone, can return: "But among any success there is much failure, much recidivism. Sometimes, you want to go on loving the pain. And then, beyond this, yet another question sharply outlines itself on the cloud: is 'success' at grief, at mourning, at sorrow, an achievement, or merely a new given condition? Because the notion of free will seems irrelevant here; the attribution of purpose and virtue—the idea of grief-work rewarded—feels misplaced." Barnes also writes of the changes in his heart and his tears: "There are moments which appear to indicate some kind of progress. When the tears—the daily, unavoidable tears—stop. When concentration returns, and a book can be read as before. When foyer-terror departs."[14]

With Roger Rosenblatt, grief after the sudden death of his adult daughter stayed for a long time. He wrote a memoir of his grief that described the depth of his anger, then he wrote another. He couldn't give up his anger. Its depth was a constant. As clinicians will sometimes say to patients: "It's easier to be mad than sad." Mad casts a thick veil over sad, both hiding and protecting the grief that is not ready to be seen.

Rosenblatt's second book is different. It ends with love, a paean of joy to his daughter: "Love conquers death. No celestial jury will bring Amy back to me. I will not see her either, no matter how others may want me to. She will not talk to me. But in the time since she died, I have been aware, every minute, of my love for her. She lives in my love. This morning when I climbed into my kayak and headed out, I knew that I would be going nowhere, as I have been going nowhere for the past two and a half years. But my love for my daughter makes somewhere out of nowhere. In this boat, on this creek, I am moving forward, even as I am moving in circles. Amy returns in my love, alive and beautiful. I have her still."[15] His grief is transduced into joy with the love he experiences in the memories of his Amy, his beloved daughter.

Auspicious Signs

In Max Porter's short novel about a wifeless father with his motherless boys, the father tells of (the imaginary) Crow who, when he arrives, announces that he will stay with the father and his two sons until he's no longer needed. Then Crow takes over the narration and describes how the wife/mother died. Her death was unpredicted: she fell, hitting her head. Crow tells the father and his sons that his stay will last until their grief isn't mucking them up the way it was when he arrived, black and a bit smelly, in a rush of wind.

Crow has a final soliloquy. It begins: "Permission to leave, I'm done," and goes on to convey more details of the woman's death. "Accident in the home./ She banged her head, dreamed a bit, was sick, slept, got up and fell,/ Lay down and died." Crow continues in his flippant yet effective voice, which sustains the boys and their father: "Connoisseurs, they were, of how to miss a mother./My absolute pleasure./Just be good and listen to birds./Long live imagined animals, the need, the capacity./Just be kind and look out for your brother." Porter has made Crow into grief with feathers, a grief who is messy, and bossy, but nurturing in his coarse language and love slaps. He's not the bird that announces death, or symbolizes death. Crow is the grief that the father and two sons suffer and endure until, as Crow says to them, "You don't need me anymore."[16]

Crows—real ones—come from the corvids, the family of oscine passerine birds that contains ravens, rooks, jays, and more. Corvids have long been symbols of death in many global religions and myths, including those of early and northern Europeans. When by military necessity the dead were left on battlefields, the scavenging nature of corvids was advantageous. It was likely through this association that people came to link these birds spiritually and artistically with death.[17]

Appearances of other birds and animals have also been said to foretell or announce death. Jennifer Holland writes of her experience in a *New York Times* opinion piece. With her mother gravely ill, and a friend maintaining her vigil for her while she takes a break, Holland drives around. Her attention is drawn to a bald eagle, a huge bird whose wings can span eight feet. It is February and the day is gray with cold, slicing winds. Holland watches the eagle perched way up on a branch in a row of trees. The creature, magnificent and unusual in suburbia, is perched there motionless. While observing the one, several other eagles appear one by one, dropping or swooping down to perch on tree branches until there are nine in all. Some of them are moving as they perch, preening and twisting, flexible despite their impossible bigness.

Holland returns home, amazed. Her mother dies later that same day.[18]

Change and Knowledge

The word "crow," and the phrase "bald eagle," trigger images in one's mind. Naming emotions, like anger, anxiety, and grief—what was only felt before being named—changes brain activity from the amygdala to the frontal cortex. To name what is unknown is the key to knowledge; naming and learning tear the mask away from mystery, neutralizing the power of what was unknown.

In the story of "Rumpelstiltskin," the "moral" (in the argot of Brothers Grimm fairy tales) is that the queen's agency, her proactive attempt to learn the goblin's name after being held hostage by the deal forced upon her, is rewarded by her liberation from the terrible bargain. She learns Rumpelstiltskin's name,

Research Holding the Torch of Knowledge. Olin L. Warner, sculptor.
Library of Congress, Prints & Photographs Division, photograph
by Carol M. Highsmith, 2007.

then uses that knowledge in a transformative way. Naming the
creature gives her the power to undo a dreadful bargain, and
keep her baby safe. As the goblin is stripped of his power, she
gains hers.

To learn, to acquire information, to obtain knowledge and
to use it to change lives is powerful. A representation of the opportunity to obtain knowledge appears on the door of the U.S.
Library of Congress. Emblazoned on the door to the right is a

bronze bas-relief by Olin L. Warner entitled *Research Holding the Torch of Knowledge*.

Knowledge illuminates, knowledge is power.

How illuminating is it, really, to mull over the metaphor of alchemy? Can grief, with its gray, leaden nature, truly become ennobled as joy and changed into gold? Isn't grief merely changed into something like itself—a different sadness or sorrow? Or can it really change into something that feels like happiness when thinking about the beloved one who died?

No formula, no prescribed steps exist. No coin will buy passage for the grief-stricken out of the realm of sorrow. No servant will find an ecstatic little man dancing around a campfire in a forest and release the bereaved. How long grief's power persists will depend on many factors, including—but not limited to—the nature of the survivor's attachment with the deceased and his or her willingness to live with the pain.

Yet to confront and brave the sorrow of a lost loved one—to grieve fully—can help move the bereft survivors toward change. As memories surface of fondness, love, and laughter, grief becomes ennobled. The loss, the experience of a loved one's death, is a kind of knowledge. Kenneth Doka and others have written that traumatic events "such as a significant loss . . . are transitional lines or turning points in our life." He continues, "The remarkable nature of human beings is that sometimes in suffering we are opened to looking at reality in new ways, that sometimes suffering engenders growth."[19] Knowing grief, understanding grief, experiencing grief can effect positive change in the bereaved. This is alchemy.

And the heart? Its beats are felt whenever a hand goes to the chest, and it no longer hurts as though it's being singed with the pain of loss. And the brain, which we cannot see or feel? We can finally say to ourselves, "Okay, I understand. It's in here." At some point, the body no longer yearns as it did—to hold, see, hear, or touch the beloved person. Everything is different. Death changed what was, and grief, having been felt, has changed *what was* to *what is.* Burnished memories continue to gleam like polished gold despite the times when a word, an aroma, or an anniversary brings a rush of intense sadness—a reminder of past loss, past grief.

It is possible for life to be enjoyed again after the death of a loved one. Mysteriously, joy and happiness return, and sadness lifts like storm clouds drifting away—at least until the weather changes, as it always can. When this happens, when reminiscences delight as well as take the breath away, when memories elicit warm chuckles as well as tears, the leaden weight of grief has alchemized into the gold of a new life.

Notes

All websites in the endnotes are current as of November 15, 2019.

Introduction

1. Charles Darwin, *The Expression of the Emotions in Man and Animals* (1872; New York: Paul Ekman, 1998).
2. Joost Smolders et al., "Tissue-Resident Memory T Cells Populate the Human Brain," *Nature Communications* 9, no. 1 (2018), doi:10.1038/s41467-018-07053-9; "Nature of Immune Cells in the Human Brain Disclosed," *Science News*, November 2, 2018, https://www.sciencedaily.com/releases/2018/11/181102083419.htm.

ONE
Evolutionary Origins

1. Antoine de Saint-Exupéry, *The Little Prince* (New York: Harcourt, 1971).
2. William Shakespeare, *Love's Labor's Lost*, act 5, scene 2, available at http://shakespeare.mit.edu/lll/full.html.
3. See the Wikipedia entry "Eleanor cross" at http://en.m.wikipedia.org/wiki/Eleanor_cross; and Sara Eliot, "The Eleanor Crosses: A Love Story in Stone," TimeTravel-Britain.com, at https://www.timetravel-britain.com/articles/history/eleanor.shtml.

4. History.com editors, "Taj Mahal," at http://www.history.com/topics/taj -mahal, updated June 6, 2019.

5. Carl Safina, *Beyond Words: What Animals Think and Feel* (New York: Picador, 2015), 72–73.

6. Barbara J. King, *How Animals Grieve* (Chicago: University of Chicago, 2013), 1–2.

7. "Orca Carries Dead Calf for Week during 'Deep Grieving,'" Associated Press, July 30, 2018, at https://www.ctvnews.ca/sci-tech/orca-carries-dead -calf-for-week-during-deep-grieving-1.4034245.

8. From Barbara J. King's personal website; see http://www.barbarajking.com /media.

9. King, *How Animals Grieve*, 94.

10. John M. Marzluff and Tony Angell, *In the Company of Crows and Ravens* (New Haven: Yale University Press, 2005), 190.

11. See the Wikipedia entry "Prairie vole," at https://en.m.wikipedia.org/wiki /Prairie_vole.

12. King, *How Animals Grieve*, 52.

13. Charles Darwin, *The Expression of the Emotions in Man and Animals* (1872; New York: Paul Ekman Group, 1998).

14. University of California Museum of Paleontology website, "The Mesozoic Era," http://www.ucmp.berkeley.edu/mesozoic/mesozoic.php.

15. David Norman, *Dinosaurs: A Very Short Introduction*, 2nd ed. (London: Oxford University Press, 2017), 39.

16. Bakker is quoted in Richard Conniff, *House of Lost Worlds: Dinosaurs, Dynasties, and the Story of Life on Earth* (New Haven: Yale University Press, 2016), 268.

17. King, *How Animals Grieve*, 102.

18. Ibid., 102.

19. Ibid., 163.

20. Safina, *Beyond Words*, 73.

21. Norman, *Dinosaurs*, 148–153; Douglas Preston, "Annals of the Former World: The Day the Dinosaurs Died; A Discovery Sheds Light on the Dinosaurs' Final Hours," *New Yorker*, April 8, 2019; "Dinosaur-Killing Asteroid Impact Cooled Earth's Climate More Than Previously Thought," ScienceDaily.com, October 31, 2017, https://www.sciencedaily.com /releases/2017/10/171031111446.htm.

22. Preston, "Annals of the Former World."

23. Dennis O'Neill, "The First Primates," available at https://www2.palomar .edu/anthro/earlyprimates/early_2.htm.

24. T. S. Kemp, *Mammals: A Very Short Introduction* (London: Oxford University Press, 2017), 50–54.

25. Bob Strauss, "The Miocene Epoch (23–5 Million Years Ago)," ThoughtCo .com, July 25, 2018, https://thoughtco.com/the-miocene-epoch-1091366.

26. Mark Jobling et al., *Human Evolutionary Genetics*, 2nd ed. (New York: Garland Science, 2014), 229.

27. Megan Y. Dennis et al., "Evolution of Human-Specific Neural SRGAP2 Genes by Incomplete Segmental Duplication," *Cell* 149, no. 4 (2012): 912–922, http://quantamagazine.org/20140102-a-missing-genetic-link-in-human-evolution; ibid., 274–275.

28. Raymond Dart, "*Australopithecus africanus:* The Man-Ape of South Africa," *Nature* 115 (1925): 195–199.

29. Carol V. Ward et al., "Thoracic Vertebral Count and Thoracolumbar Transition in *Australopithecus afarensis,*" *Proceedings of the National Academy of Science* 114, no. 23 (2017): 6000–6004; Peter Holley, "How a 3.3-Million-Year-Old Toddler Offers Researchers a Window into Human Evolution," *Washington Post*, May 26, 2017.

30. Lewis S. B. Leakey et al., "A New Species of the Genus *Homo* from Olduvai Gorge," *Nature* 202 (1964): 7–9.

31. Sonia Harmand et al., "3.3-Million-Year-Old Stone Tools from Lomekwi 3, West Turkana, Kenya," *Nature* 521 (2015): 310–315, doi:10.1038/nature 14464.

32. Ibid.

33. Megan Y. Dennis et al., "Evolution of Human-Specific Neural SRGAP2 Genes by Incomplete Segmental Duplication," *Cell* 149, no. 4 (2012): 912–922; and ibid., citing Jobling et al., *Human Evolutionary Genetics* (New York: Garland, 2004). According to Dennis et al., the first duplication of the SRGAP2 gene was a partial one. It occurred around 3.4 million years ago and resulted in SRGAP2B. About a million years later, the second larger duplication occurred and produced SRGAP2C.

34. Dennis et al., "Evolution," 920.

35. Lee R. Berger and Ron J. Clarke, "Eagle Involvement in Accumulation of Taung Child Fauna," *Journal of Human Evolution* 29 (1995): 275–299; Lee R. Berger, "Brief Communication: Predatory Bird Damage to the Taung Type-Skull of *Australopithecus africanus* Dart 1925," *American Journal of Physical Anthropology* 131 (2006): 166–168.

36. Dora Biro et al., "Chimpanzee Mothers at Bossou, Guinea Carry the Mummified Remains of Their Dead Infants," *Current Biology* 20 (2010): R351–R352; King, *How Animals Grieve*, 81–83.

37. James R. Anderson, Alasdair Gilles, and Louise C. Lock, "Pan Thanatology," *Current Biology* 20 (2010): R349–R351.

38. Brian Switek, "What Death Means to Primates," Wired.com, April 18, 2011, http://www.wired.com/2011/04/what-death-means-to-primates.

39. Paul Pettitt, *The Paleolithic Origins of Human Burial* (London: Routledge, 2011), 7.

40. Enrico de Lazaro, "Sima de los Huesos Hominins Were Actually Early Neanderthals, Say Anthropologists," www.Sci-News.com, March 15, 2016,

http://www.sci-news.com/othersciences/anthropology/sima-de-los-huesos
-hominins-early-neanderthals-03703.html.

41. Pettitt, *Paleolithic Origins*, 9.
42. Ibid., 262. Erella Hovers and Anna Belfer-Cohen note, "Burial is perceived as one of the human cultural reactions to the phenomenon of death." See Hovers and Belfer-Cohen, "Insights into Early Mortuary Practices of *Homo*," in *The Oxford Handbook of the Archaeology of Death and Burial*, ed. Sarah Tarlow and Liv Nilsson Stutz (London: Oxford University, 2013), 631.
43. Ibid., 634.
44. Julien Riel-Salvatore and Claudine Gravel-Miguel, "Upper Paleolithic Mortuary Practices in Eurasia," in *The Oxford Handbook of the Archaeology of Death and Burial*, ed. Sarah Tarlow and Liv Nilsson Stutz (London: Oxford University, 2013), 329.
45. Vincenzo Fomicola and Alexandra P. Buzhilova, "Double Child Burial from the Sunghir (Russia): Pathology and Inferences for Upper Paleolithic Funerary Practices," *American Journal of Physical Anthropology* 124 (2004): 189–198.
46. Ibid.
47. Caitlin Doughty, *From Here to Eternity* (New York: Norton, 2017), 75; Amanda Bennett, "When Death Doesn't Mean Goodbye," *National Geographic* (March 2016), https://www.nationalgeographic.com/magazine/2016/04/death-dying-grief-funeral-ceremony-corpse.
48. See the Wikipedia entry "Qingming Festival," at https://en.wikipedia.org/wiki/Qingming_Festival.
49. See the Wikipedia entry "Memorial Day," at https://en.m.wikipedia.org/wiki/Memorial_Day.
50. Personal communication from an acquaintance describing an Eastern Orthodox funeral.
51. Safina, *Beyond Words*, 30.
52. To consider this powerful emotion from an evolutionary perspective can help us picture grief in the dinosaurs, those ancient creatures that John H. Ostrom made "real and visceral," first to his colleagues in paleontology, and later to the world. See Richard Conniff, *House of Lost Worlds*, 272, quotations from John N. Wilford, *New York Times*, July 21, 2005, and from the *Sunday Times* (London), August 9, 2005.

TWO

Forms of Grief

1. See the Online Etymology Dictionary entry "attach," at https://www.etymonline.com/search?q=attach.

2.	John Archer, *The Nature of Grief: The Evolution and Psychology of Reactions to Loss* (London: Routledge, 2001), 48.

3.	Donald Winnicott, *Home Is Where We Start From* (New York: Norton, 1990), 119–120, 144.

4.	Ibid., 33, 65.

5.	John Evelyn, *Sylva; or, A Discourse of Forest-Trees and the Propagation of Timber in His Majesty's Dominions* (London: Royal, 1664), 133, and (York, 1801), 3d ed., vol. 1, 272, https://pegsandtails.wordpress.com/2011/12/25/the-common-holly/#_edn2.

6.	Pauline Boss, Susan Roos, and Darcy L. Harris, "Grief in the Midst of Ambiguity and Uncertainty," in Robert A. Neimeyer, Darcey L. Harris, Howard R. Winokuer, and Gordon F. Thornton, eds., *Grief and Bereavement in Contemporary Society* (New York: Routledge, 2011), 163–175.

7.	Ibid., 166.

8.	Erich Lindemann, "Symptomatology and Management of Acute Grief," *American Journal of Psychiatry* 101, no. 2 (1944): 141–148, doi:10.1176/ajp.101.2.141.

9.	Kenneth J. Doka, *Grief Is a Journey: Finding Your Path through Loss* (New York: Atria, 2016), 187–193.

10.	Ibid., 170–181.

11.	Lindemann, "Symptomatology and Management of Acute Grief."

12.	George Bonanno, *The Other Side of Sadness* (New York: Basic Books, 2010), 6–9; George Bonanno et al., "Psychological Resilience after Disaster: New York City in the Aftermath of the September 11th Terrorist Attack," *Psychological Science* 17, no. 3 (March 2006): 181–186; Archer, *Nature of Grief*, 114.

13.	Bonanno, *Other Side of Sadness*, 6; Archer, *Nature of Grief*, 218.

14.	J. William Worden, *Grief Counseling and Grief Therapy*, 5th ed. (New York: Springer, 2018), 143–145, 160; Archer, *Nature of Grief*, 114.

15.	Archer, *Nature of Grief*, 191.

16.	Judith L. M. McCoyd and Carolyn Ambler Walter, *Grief and Loss across the Lifespan: A Biosocial Perspective* (New York: Springer, 2016), 99.

17.	Archer, *Nature of Grief*, 53, 114, 125.

18.	Helene Deutsch, "Absence of Grief," *Psychoanalytic Quarterly* (1937), 6: 12–22.

19.	M. Katherine Shear, "Grief and Mourning Gone Awry: Pathway and Course of Complicated Grief," *Dialogues in Clinical Neuroscience* 14 (2015): 119–129; M. Katherine Shear, "Complicated Grief," *New England Journal of Medicine* 372 (2015): 153–160, doi:1056/NEJMr13115618; for 10 percent, see M. Katherine Shear, "Complicated Grief and Related Bereavement Issues for DSM-5," *Depression and Anxiety* 28, no. 2 (February 2011): 103–117, doi:10.1002/da.20780.

20.	"Number of Deaths in the United States, 2015," Centers for Disease

Control and Prevention website, https://www.cdc.gov/nchs/data/databriefs/db267.pdf.

21. Shear, "Complicated Grief."

22. Archer, *Nature of Grief*, 160.

23. Worden, *Grief Counseling*, 143–145, 160.

24. Ibid., 147–149.

25. Archer, *Nature of Grief*, 225–226.

26. Ibid., 112. Archer cites L. G. Peppers and J. Knapp, who write that there is a "lack of opportunity to express perinatal grief to sympathetic listeners." L. G. Peppers and J. Knapp, *Motherhood and Mourning* (New York: Praeger, 1980).

27. Lonneke I. M. Lenferink et al., "Reciprocal Associations among Symptom Levels of Disturbed Grief, Posttraumatic Stress, and Depression Following Traumatic Loss: A Four-Wave Cross-Lagged Study," *Clinical Psychological Science* 1–10 (2019), doi: 10.1177/2167702619858288; G. E. Smid et al., "Brief Electric Psychotherapy for Traumatic Grief (BEP-TG): Toward Integrated Treatment of Symptoms Related to Traumatic Loss," *European Journal of Psychotraumatology* 6 (2015): 1–11.

28. *Diagnostic and Statistical Manual of Mental Disorders*, 5th ed. (Washington: American Psychiatric Association, 2013), 789–791; Madelyn R. Franklin and Donald J. Robinaugh, "Grief and Post-Traumatic Stress Following Bereavement," in *Clinical Handbook of Bereavement and Grief Reactions*, ed. Eric Bui (Basel, Switz.: Springer Nature, 2018), 19–44.

29. Worden, *Grief Counseling*, 110, 240.

30. Julian Barnes quotes E. M. Forster in *Levels of Life* (New York: Knopf, 2013), 75.

31. Guy Gugliotta, "New Estimate Raises Civil War Death Toll," *New York Times*, April 2, 2012, https://www.nytimes.com/2012/04/03/science/civil-war-toll-up-by-20-percent-in-new-estimate.html.

32. "Number of Deaths."

33. See the Wikipedia entry "Remembrance poppy," at https://en.m.wikipedia.org/wiki/Remembrance_poppy.

34. Quoted in Laura Collins-Hughes, "A Year of Mourning Steeped in African Tradition Leads to a Theater Piece," *New York Times*, December 31, 2015, https://www.nytimes.com/2016/01/01/theater/a-year-of-mourning-steeped-in-african-tradition-leads-to-a-theater-piece.html.

35. Sigmund Freud, "Mourning and Melancholia," in J. Strachey, ed. and trans., *The Standard Edition of the Complete Psychological Works of Sigmund Freud*, vol. 14 (London: Hogarth, 1917), 239–260.

36. Darian Leader, *The New Black: Mourning, Melancholia, and Depression* (New York: Penguin, 2009), 8.

37. Elisabeth Kübler-Ross, *On Death & Dying* (New York: Scribner, 1967).

38. Worden, *Grief Counseling*, 39.

39. Ruth Davis Konigsberg, *The Truth about Grief: The Myth of Its Five Stages and the New Science of Loss* (New York, 2011), 1–3, 11–12, 68–69, 74.

40. Bonanno, *Other Side of Sadness*, 1–9; Bonanno et al., "Psychological Resilience."

41. M. Stroebe and H. Schut, "The Dual Process Model of Coping with Bereavement: Rationale and Description," *Death Studies* 23, no. 3 (April–May 1999): 197–224, doi:10.1080/074811899201046.

42. Worden, *Grief Counseling*, 41–53.

THREE
Language of the Bereaved

1. Stephen R. Anderson, "How Many Languages Are There in the World?" Linguistic Society of America, https://www.linguisticsociety.org/content /how-many-languages-are-there-world.

2. Osip Mandelstam, "The Swallow," in *Complete Poetry of Osip Emilevich Mandelstam*, trans. Burton Raffel and Alla Burago, introduction and notes by Sidney Monas (Albany: State University of New York Press, 1973).

3. Jesse Sheidlower, "Word Count: Are There Really 988,968 Words in the English Language?" *Slate*, April 10, 2006, https://slate.com/human-interest /2006/04/how-many-words-are-there-in-english.html.

4. Lev Semenovich Vygotsky, *Thought and Language*, trans. and ed. Eugenia Hanfmann and Gertrude Vakar (Cambridge, MA: MIT Press, 1962), 153.

5. See the Online Etymology Dictionary entry for "bereft" at https://www .etymonline.com/search?q=bereft.

6. Julian Barnes, *Levels of Life* (New York: Alfred A. Knopf, 2013), 96.

7. Percy Bysshe Shelley, "To a Skylark," in *Prometheus Unbound* (London: C. and J. Ollier, 1820).

8. Elizabeth A. Phelps and Joseph E. LeDoux, "Contributions of the Amygdala to Emotion Processing: From Animal Models to Human Behavior," *Neuron* 48 (2005): 175–187.

9. Helen Macdonald, *H Is for Hawk* (London: Jonathan Cape, 2014), 12–13.

10. Jean Rhys, *Wide Sargasso Sea* (New York: Norton, 1982), 17.

11. Terry L. Martin and Kenneth J. Doka, "The Influence of Gender and Socialization on Grieving Styles," in Robert A. Neimeyer, Darcey L. Harris, Howard R. Winokuer, and Gordon F. Thornton, eds., *Grief and Bereavement in Contemporary Society* (New York: Routledge, 2011), 70.

12. Paul C. Holinger, *What Babies Say before They Can Talk: The Nine Signals Infants Use to Express Their Feelings* (New York: Simon and Schuster, 2003).

13. Adam Phillips, *The Beast in the Nursery* (New York: Pantheon, 1998), 42.

14. Antoine Leiris, *You Will Not Have My Hate* (New York: Penguin, 2016), 46.

15. Besell Van der Kolk, *The Body Keeps the Score* (New York: Viking, 2014), 43–44.

16. Leiris, *You Will Not Have My Hate*, 97.
17. The quotations in this and the subsequent two paragraphs are from Sonia Purnell, *Clementine* (New York: Viking, 2015), 147–149.
18. Phillips, *Beast in the Nursery*, 43–44, 52–53.
19. Matthew D. Lieberman et al., "Putting Feelings into Words: Affect Labeling Disrupts Amygdala Activity to Affective Stimuli," *Psychological Science* 18 (2007): 421–428, doi: 10.1111/j.1467–9280.2007.01916.
20. Jared B. Torre and Matthew D. Lieberman, "Putting Feelings into Words: Affect Labeling as Implicit Emotion Regulation," *Emotion Review* 10, no. 2 (2018): 116–124.
21. Rebecca Soffer and Gabrielle Birkner, "How to Speak Grief," *New York Times*, January 14, 2018.
22. Stig Abel, "This Week," *Times Literary Supplement* (London), June 2, 2017, 3, http//www.the-tls.co.uk.
23. "Cussac Cave—Grotte de Cussac," DonsMaps.com, http://donsmaps.com/cussac.html.
24. Paul Pettitt, *Paleolithic Origins of Human Burial* (London: Routledge, 2011), 153.
25. C. S. Lewis, *A Grief Observed* (Greenwich, CT: Seabury, 1963), 56.
26. Archer, *Nature of Grief*, 206–208. Archer cites A. N. Wilson's 1990 biography of C. S. Lewis.
27. Meghan O'Rourke, *The Long Goodbye: A Memoir* (New York: Riverhead, 2011), 171.
28. Soffer and Birkner, "How to Speak Grief."
29. Barnes, *Levels of Life*, 80, 111.
30. Ibid., 80.
31. Sheryl Sandberg and Adam Grant, "How to Talk to a Loved One Who Is Suffering," and Belinda Luscombe, "Let's Talk about Grief," both in *Time*, April 24, 2017.
32. Max Porter, *Grief Is the Thing with Feathers* (Minneapolis: Graywolf, 2015), 6–8.
33. Jane Brody, "Coloring Your Way through Grief," *New York Times*, May 16, 2016.
34. Louise Glück, "The Wild Iris," in Glück, *The Wild Iris* (New York: Harper-Collins, 1992), 1.
35. Paul Pettitt, *Paleolithic Origins of Human Burial* (London: Routledge, 2011), 35.

FOUR

The Grief-Stricken Brain

1. The evocative phrase "memory pools" comes from Joyce Carol Oates, *A Widow's Story: A Memoir* (New York: Ecco, 2012), 46.

2. Bryan Kolb and Ian Q. Wishaw, *An Introduction to Brain and Behavior*, 4th ed. (New York: Worth, 2014), 37, 42; see the Wikipedia entry "Pia mater," at https://en.m.wikipedia.org/wiki/Pia_mater.

3. See the Wikipedia entry "Scientia potential est," at https://en.wikipedia.org/wiki/Scientia_potentia_est. The phrase has been attributed to Sir Francis Bacon, who refers to "ipsa scientia potestas est" (knowledge itself is power), in his *Meditationes Sacrae* (1597).

4. Barbara Leaming, *Jacqueline Bouvier Kennedy Onassis: The Untold Story* (New York: St. Martin's, 2014), 132–134.

5. Ibid., 126.

6. Ibid., 266, 293–294.

7. *Diagnostic and Statistical Manual of Mental Disorders*, 5th ed. (Washington: American Psychiatric Association, 2013), 271–280, describes PTSD as a traumatic condition experienced by veterans and others, due to an "exposure to actual or threatened death, serious injury, or sexual violence," 271.

8. Alex Korb, *The Upward Spiral* (Oakland, CA: New Harbinger, 2015), 18–19.

9. Maximilien Chaumon, Kestutis Kveraga, Lisa Feldman Barrett, and Moshe Bar, "Visual Predictions in the Orbitofrontal Cortex Rely on Associative Content," *Cerebral Cortex* 24, no. 11 (2014): 2899–2907. Human neuroimaging studies have provided evidence of face-selective areas in the orbitofrontal cortex of the prefrontal region. See Winrich Freiwald, Bradley Duchaine, and Galit Yovel, "Face Processing Systems: From Neurons to Real World Social Perception," *Annual Review of Neuroscience* 39 (2016): 325–346, doi:10.1146/annurev-neuro-070815-013934; Sid Gilman and Sarah Winans Newman, eds., *Manter and Gatz's Essentials of Clinical Neuroanatomy and Neurophysiology*, 10th ed. (Philadelphia: F. A. Davis, 2003), 188.

10. Gilman and Newman, *Manter and Gatz's Essentials*, 188; Korb, *Upward Spiral*, 19.

11. Korb, *Upward Spiral*, 20.

12. Gilman and Newman, *Manter and Gatz's Essentials*, 193.

13. Ibid., 193; David L. Clark, Nashaat N. Boutros, and Mario F. Mendez, *The Brain and Behavior: An Introduction to Behavioral Neuroanatomy* (New York: Cambridge University Press, 2006), 223.

14. L. L. Bruce and T. J. Leary, "The Limbic System of Tetrapods: A Comparative Analysis of Cortical and Amygdalar Populations," *Brain Behavior Evolution* 46, nos. 4–5 (1995): 224–234, doi: 10.1159/000113276.

15. Most structures within the brain are bilateral, so there are two amygdalas, buried in the right and left temporal lobes of the brain.

16. Elizabeth A. Phelps and Joseph E. LeDoux, "Contributions of the Amygdala to Emotion Processing: From Animal Models to Human Behavior," *Neuron* 48 (2005): 175–187; see the Wikipedia entry "Ventromedial prefrontal cortex," at http://en.m.wikipedia.org/wiki/Ventromedial _prefrontal_cortex.

17. Gilman and Newman, *Manter and Gatz's Essentials*, 189, 195.

18. Eleanor A. Maguire et al., "Navigation-Related Structural Change in the Hippocampi of Taxi Drivers," *PNAS* 97, no. 8 (2000): 4398–4403, http://m.pnas.org/content/97/8/4398.full.

19. Gerd Kempermann et al., "Human Adult Neurogenesis: Evidence and Remaining Questions," *Cell Stem Cell* (July 5, 2018), 23.

20. C. Cooper, H. Y. Moon, and Henrietta van Praag, "On the Run for Hippocampal Plasticity," *Cold Spring Harbor Perspectives in Medicine* (May 11, 2017), pii:a029736, doi: 10.1101/cshperspect.a029736 (epub ahead of print). See also Gretchen Reynolds, "Get Moving, for Your Mind's Sake," *New York Times*, October 10, 2017.

21. Kolb and Wishaw, *Introduction to Brain and Behavior*, 54.

22. Esther M. Sternberg, *The Balance Within: The Science Connecting Health and Emotions* (New York: W. H. Freeman, 2001), 58.

23. Robert A. McGovern and Sameer A. Sheth, "Role of the Dorsal Anterior Cingulate Cortex in Obsessive-Compulsive Disorder: Converging Evidence from Cognitive Neuroscience and Psychiatric Neurosurgery," *Journal of Neurosurgery* (2017), 126, 132–147. Lisa M. Shin et al., "Dorsal Anterior Cingulate Function in Posttraumatic Stress Disorder," *Journal of Traumatic Stress* 20, no. 5 (2007): 701–712.

24. Harald Gundel et al., "Functional Neuroanatomy of Grief: An fMRI Study," *American Journal of Psychiatry* 160 (2003): 1946–1953, doi:10.1176/appi.ajp.160.11.1956.

25. Ibid.

26. M. Katherine Shear, "Grief and Mourning Gone Awry: Pathway and Course of Complicated Grief," *New Dialogues in Clinical Neuroscience* 14, no. 2 (June 2012): 119–128; M. Katherine Shear, "Complicated Grief," *New England Journal of Medicine* 372 (2015): 153–160, doi: 10.1056/NEJMcp1315618; M. Katherine Shear, "Complicated Grief and Related Bereavement Issues for DSM-5," *Depression and Anxiety* 28, no. 2 (February 2011): 103–117, doi:10.1002/da.20780.

27. Mary-Francis O'Connor et al., "Craving Love? Enduring Grief Activates Brain's Reward Center," *Neuroimage* 42 (2008): 969–972, doi:10.1016/j.neuroimage.2008.04.256appi.ajp.160.11.1956.

28. John Dowling, *Understanding the Brain* (New York: W.W. Norton, 2018), 245–247.

29. Mary Frances O'Connor and Mairead H. McConnell, "Grief Reactions: A Neurobiological Approach," in *Clinical Handbook of Bereavement and Grief Reactions*, ed. Eric Bui (Basel, Switz.: Springer Nature, 2018), 45–62.

30. Winrich A. Freiwald, Brad Duchaine, and Galit Yovel, "Face Processing Systems: From Neurons to Real-World Social Perception," *Annual Review of Neuroscience* 39 (2015): 325–346. See also Dowling, *Understanding the*

Brain, 160–161: "Exactly what is involved in recognizing faces is not well understood."

31. Josef Parvizi et al., "Electrical Stimulation of Human Fusiform Face-Selective Regions Distorts Face Perception," *Journal of Neuroscience* 32, no. 43 (October 2012): 14915–14920.

32. Benjamin Libet, Curtis A. Gleason, Elwood W. Wright, and Dennis K. Pearl, "Time of Conscious Intention to Act in Relation to Onset of Cerebral Activity (Readiness Potential): The Unconscious Initiation of A Freely Voluntary Act," *Brain* 106, no. 3 (September 1, 1983): 623–642.

33. Mark Hallett, "Physiology of Free Will," *Annals of Neurology* 80 (2016): 5–12, doi:10.1002/ana.24657.

34. Stanislas Dehaene, *Consciousness and the Brain: Deciphering How the Brain Codes Our Thoughts* (New York: Penguin, 2014), 53.

35. Peter Wohlleben, *The Hidden Life of Trees* (Vancouver: Greystone Books, 2015), 49–51.

FIVE

The Broken Heart of Grief

1. Elaine A. Evans, "Ancient Egypt: The Sacred Scarab," McClung Museum online, April 17, 1996, at http://museum.unl.edu/research/entomology /Egyptian_Sacred_Scarab/egs-text.htm; see the Britannica.com entry for "Ib" at https://www.britannica.com/topic/ib.

2. Sandeep Jauhar, *Heart: A History* (New York: Farrar, Straus and Giroux, 2018), 18, 36–37.

3. Carl Zimmer, "A Speedier Way to Catalogue Human Cells," *New York Times*, August 22, 2017, D6.

4. Jauhar, *Heart*, 46.

5. Ibid., 46–47.

6. Maylis de Kerangal, *Mend the Living*, trans. Jessica Moore (London: Maclehose Press, 2016), 226.

7. National Heart, Lung, and Blood Institute, "Broken Heart Syndrome," at https://www.nhlbi.nih.gov/health/health-topics/topics/hhw/broken-heart -syndrome/other-names.

8. Jauhar, *Heart*, 24.

9. "Takotsubo Cardiomyopathy: Broken-Heart Syndrome," Harvard Health Publishing, updated April 2, 2018, at https://www.health.harvard.edu/heart -health/takotsubo-cardiomyopathy-broken-heart-syndrome.

10. Jauhar, *Heart*, 25.

11. Ibid., 24; Anahad O'Connor, "Your Heart Has Feelings, Too," *New York Times*, November 6, 2018, https://www.nytimes.com/2018/10/30/well/live /how-emotions-can-affect-the-heart.html.

12. "Takotsubo Cardiomyopathy."
13. Sylvia J. Buchmann, Dana Lehmann, and Christin E. Stevens, "Takotsubo Cardiomyopathy—Acute Cardiac Dysfunction Associated with Neurological and Psychiatric Disorders," *Frontiers in Neurology* 10 (August 2019): 1–8, doi:10.3389/fneur.2019.00917.
14. Elizabeth Mostofsy et al., "Risk of Acute Myocardial Infarction after the Death of a Significant Person in One's Life," *Circulation* 125 (2012): 491–496, doi:10.1161/circulationaha.111.061990.
15. Glenn Ringtved, *Cry, Heart, But Never Break*, trans. Robert Moulthrop, illus. Charlotte Pardi (1978; New York: Enchanted Lion Books, 2016).
16. "Cardiac Ablation Procedures," MedlinePlus.gov, https://medlineplus.gov/ency/article/007368.htm.
17. William Shakespeare, *Romeo and Juliet*, in A. L. Rowse, *Annotated Shakespeare* (New York: Clarkson N. Potter, 1978), 3:132–133.
18. De Kerangal, *Mend the Living*, 138–141.
19. Ibid., 227–229.
20. See the Center for Organ Recovery and Education at https://core.org. The true-life story described in this and the following paragraphs can be found in Katie Rogers, "Guided down the Aisle by Her Father's Heart," *New York Times*, January 1, 2017.
21. Sarah DiGiulio, "'Beating Heart in a Box' Promises Major Revolution in Medical Care," NBCNews.com, https://www.nbcnews.com/mach/technology/beating-heart-box-promises-major-revolution-medical-care-n770236.

SIX

The Grieving Body

1. Meghan O'Rourke, *The Long Goodbye* (New York: Riverhead, 2011), 125–126.
2. Oliver Sacks, *Hallucinations* (New York: Vintage, 2012), 230.
3. Ibid., 231.
4. Darian Leader, *The New Black: Mourning, Melancholia and Depression* (London: Hamish Hamilton, 2008), 201.
5. See the Wikipedia entry "Eros (concept)" at https://en.m.wikipedia.org/wiki/Eros_(concept).
6. See Washington Irving's quotation at http://quotegeek.com/literature/washington-irving/5895.
7. Martin Meredith, *Elephant Destiny: Biography of an Endangered Species* (New York: PublicAffairs, 2003), 115, https://www.elephantsforever.co.za/elephant-emotions-grieving.html; http://scienceline.ucsb.edu/getkey.php?key=1067.
8. See the Wikipedia entry "Tears," at https://en.m.wikipedia.org/wiki/Tears;

Reena Mukamal, "All about Emotional Tears," American Academy of Ophthalmology website, Aao.org, February 28, 2017.

9. See the Wikipedia entry "Ella Freeman Sharpe" at https://en.m.wikipedia .org/wiki/Ella_Freeman_Sharpe.

10. Quoted in Ella Freeman Sharpe, *Dream Analysis* (New York: Brunner /Mazel: 1978), 176.

11. Ibid., 176. Sharpe's patient was quoting "In Memoriam," by Alfred Lord Tennyson.

12. William H. Frey, foreword to Rose-Lynn Fisher, *The Topography of Tears* (New York: Bellevue Literary Press, 2017), 9–10. See also William H. Frey and Muriel Langseth, *Crying: The Mystery of Tears* (Minneapolis: Winston Press, 1985).

13. Frey, foreword to Fisher, *Topography of Tears*, 8.

14. Fisher, *Topography of Tears*, 47, 19, 45.

15. Joost Smolders et al., "Tissue-Resident Memory T Cells Populate the Human Brain," *Nature Communications* 9, no. 1 (2018), doi:10.1038s41467–018–07053–9; "Nature of Immune Cells in the Human Brain Disclosed," ScienceDaily.com, November 2, 2018, https://science daily.com/releases/2018/11/181102083419.htm.

16. Antoine Louveau et al., "Structural and Functional Features of Central Nervous System Lymphatic Vessels," *Nature* 523 (2015): 337–341; Jeffrey J. Iliff, Steven A. Goldman, and Maiken Nedergaard, "Implications of the Discovery of Brain Lymphatic Pathways," *Lancet Neurology* 14, no. 10 (2015): 977–979.

17. Michal Schwartz, *Neuroimmunity* (New Haven: Yale University Press, 2015), 39; Anthony J. Filiano, Sachin P. Gadani, and Jonathan Kipnis, "How and Why Do T Cells and Their Derived Cytokines Affect the Injured and Healthy Brain?" *Nature Reviews/Neuroscience* 18 (June 2015): 375–384.

18. Filiano, Gadani, and Kipnis, "How and Why Do T Cells and Their Derived Cytokines Affect the Injured and Healthy Brain?"

19. Bryan Kolb and Ian Q. Whishaw, *Fundamentals of Human Neuropsychology* (New York: Worth, 2009), 721–723.

20. J. William Worden, *Grief Counseling and Grief Therapy* (New York: Springer, 2018), 147–149.

21. Ibid., 148. (Worden cites the term used by Zisook and DeVaul to describe physical symptoms experienced by the survivor that are similar to those of the deceased.)

SEVEN
Mothers

1. Scott Simon, "Tweeting Mom's Goodbye," *New York Times*, March 29, 2015, https://opinionator.blogs.nytimes.com/author/scott-simon.

2. Ibid.
3. Meghan O'Rourke, *The Long Goodbye: A Memoir* (New York: Riverhead, 2011), 76.
4. Roger Rosenblatt, *Making Toast* (New York: HarperCollins, 2010), 115.
5. Ibid., 82.
6. O'Rourke, *Long Goodbye*, 171. Emphasis in the original.
7. Julian Barnes, *Levels of Life* (New York: Knopf, 2013), 127.
8. William Shakespeare, *Julius Caesar*, ed. Barbara A. Mowat and Paul Werstine (New York: Simon & Shuster, 2011), 13, 15.

EIGHT
Fathers

1. See the Wikipedia entry *"Oedipus Rex"* at https://en.m.wikipedia.org/wiki /Oedipus_Rex.
2. Peter Gay, *Freud: A Life for Our Time* (New York: W. W. Norton, 1988), 100.
3. Adam Phillips, *Becoming Freud: The Making of a Psychoanalyst* (New Haven: Yale University Press, 2014), 104.
4. Ibid., 104–105.
5. Ibid., 26.
6. Saul McLeod, "What Are the Most Interesting Ideas of Sigmund Freud?," SimplyPsychology.org, updated 2018, https://www.simplypsychology.org /Sigmund-Freud.html; the Wikipedia entry "Psychoanalysis" at https:// en.wikipedia.org/wiki/Psychoanalysis; and Sigmund Freud, "The Interpretation of Dreams," in *The Standard Edition of the Complete Psychological Works of Sigmund Freud* (London: Hogarth Press, 1956–1974).
7. Matthew D. Lieberman et al., "Putting Feelings into Words: Affect Labeling Disrupts Amygdala Activity to Affective Stimuli," *Psychological Science* 18 (2007): 421–428; Jared B. Torre and Matthew D. Lieberman, "Putting Feelings into Words: Affect Labeling as Implicit Emotion Regulation," *Emotion Review* 10, no. 2 (April 2018): 116–124, https://journals .sagepub.com/doi/pdf/10.1177/2472555217701685.
8. Richard Alleyne, "Writing Poems Helps Brains Cope with Emotional Turmoil, Say Scientists," *Telegraph* (London) website, https://www.tele graph.co.uk/culture/culturenews/4630043/AAAS-Writing-poems-helps -brain-cope-with-emotional-turmoil-say-scientists.html.
9. See the Wikipedia entry "Electra complex" at https://en.wikipedia.org /wiki/Electra_complex. Sigmund Freud, in contrast to Carl Jung, called the girl's conflict with her mother for her father's affection a feminine Oedipus attitude.
10. Gay, *Freud*, 390.

11. Patsy Rodenburg, *The Actor Speaks: Voice and the Performer* (New York: St. Martin's, 2000), 108–110.

12. Pew Research Center, "The American Family Today," December 17, 2015, http://www.pewsocialtrends.org/2015/12/17/1-the-american-family -today.

13. Luigi Zoja, *The Father: Historical, Psychological and Cultural Perspectives*, rev. ed., trans. Henry Martin (London: Routledge, 2018), xv.

14. Ibid., 15.

15. See the Wikipedia entry "Falconry," at https://en.m.wikipedia.org/wiki /Falconry#Husbandry.2C_and_equipment.

16. Helen Macdonald, *H Is for Hawk* (London: Jonathan Cape, 2014), 59.

17. Ibid., 220.

18. Clea Simon, *Fatherless Women: How We Change after We Lose Our Dads* (New York: John Wiley & Sons, 2002), 107.

19. Macdonald, *H Is for Hawk*, 57.

20. Darian Leader, *The New Black: Mourning, Melancholia and Depression* (London: Hamish Hamilton, 2008), 49–50.

NINE

Children

1. See the Wikipedia entry "Little Boy Blue," at https://en.m.wikipedia.org /wiki/Little_Boy_Blue_(poem).

2. Eugene Field, *A Little Book of Western Verse* (New York: Scribner, 1889), 8–9.

3. Ibid., 130–132. Carol Rumens, a contemporary poet who writes for the UK *Guardian*, once made "Wynken, Blynken and Nod" her weekly pick and provides the following rationale for choosing it: "The poem is a virtuoso piece of rhythmic structuring, with its tripping four-beat/three-beat lines, and the master strokes of variation which extend the second quatrain of each stanza, adding a fifth rhymed line, and dividing the last into monometers. The separation insists that each name is fully relished. It turns the print on the page into a voice. This child's poetic lullaby simultaneously stimulates the imagination and soothes with its delicious sound and rhythm." Carol Rumens, *The Guardian*, March 26, 2012, https://www.theguardian .com/books/booksblog/2012/mar/26/poem-week-wynken-blynken-nod.

4. Field, *Little Book of Western Verse*, 8.

5. "Family Life," Darwin Project, at https://www.darwinproject.ac.uk/people /about-darwin/family-life/death-anne-elizabeth-darwin.

6. Darwin Correspondence Project, "Letter no. 1425," http://www.darwin project.ac.uk/DCP-LETT-1425.

7. Nate Chinen, "A Wrenching Grief Assuages with Beauty," *New York Times*,

November 27, 2014, C1–7, https://www.nytimes.com/2014/11/27/arts
/music/jimmy-greenes-beautiful-life-is-a-eulogy-to-a-daughter.html.

8. Ibid. The article includes a link to the video of Ana and Isiah singing.
9. Sukey Forbes, *The Angel in My Pocket* (New York: Penguin, 2014), 25–26, 70.
10. Ibid., 122.
11. Ibid., 162–163.
12. T. H. White, *The Once and Future King* (London: Collins, 1958).
13. "How to Choose a Youth Archery Kids Bow and Arrow Set," 3riversarchery,
https://www.3riversarchery.com/blog/youth-archery-bow-kits-for-kids.
14. Michael Paulson, "'Hamilton' and Heartache: Living the Unimaginable,"
New York Times, November 16, 2016, https://www.nytimes.com/2016/10
/16/theater/oskar-eustis-public-theater.html.
15. J. William Worden, *Grief Counseling and Grief Therapy*, 5th ed. (New York:
Springer, 2018), 185–188.
16. Roger Rosenblatt, *Making Toast* (New York: HarperCollins, 2010), 2–3.
17. "What Is Anomalous Coronary Artery?" StanfordHealthCare.org, https://
stanfordhealthcare.org/medical-conditions/blood-heart-circulation
/anomalous-coronary-artery.html.
18. Rosenblatt, *Making Toast*, 3.
19. Ibid., 2.
20. Edward Hirsch, *Gabriel: A Poem* (New York: Knopf, 2016).
21. Emily Rapp, "Expression of Grief," *New York Times Sunday Book Review*,
November 9, 2014, 13.
22. The quotations in this paragraph and the next are from Paula Span, "A
Child's Death Brings 'Trauma That Doesn't Go Away,'" *New York Times*,
October 2, 2017, https://www.nytimes.com/2017/09/29/health/children
-death-elderly-grief.html.
23. Edward Hirsch, "*Falling Out of Time* by David Grossman," *New York Times
Sunday Book Review*, July 18, 2014, https://www.nytimes.com/2014/07/20
/books/review/falling-out-of-time-by-david-grossman.html.
24. David Grossman, *Falling Out of Time*, trans. Jessica Cohen (New York:
International Vintage, 2014), 3–4, 189, 192, 30.

TEN
Sisters and Brothers

1. Kenneth J. Doka, *Grief Is a Journey* (New York: Atria, 2016), 173.
2. Ibid., 173–175.
3. J. William Worden, *Grief Counseling and Grief Therapy* (New York: Springer,
2018), 4.
4. David W. Kissane and Nadine A. Kasparian, "Theoretical Models Guiding
Our Understanding of Sibling Bereavement," in Brenda J. Marshall and

Howard R. Winokeur, eds., *Sibling Loss across the Lifespan* (London: Routledge, 2017), 11.

5. John Archer, *The Nature of Grief: The Evolution and Psychology of Reactions to Loss* (London: Routledge, 2001), 217–219.

6. Darwin Correspondence Project, "The Death of Anne Elizabeth Darwin," https://www.darwinproject.ac.uk/people/about-darwin/family-life/death-anne-elizabeth-darwin.

7. Sukey Forbes, *The Angel in My Pocket* (New York: Viking, 2013), 79.

8. Darwin Correspondence Project, "To W. D. Fox, 29 April [1851]," http://www.darwinproject.ac.uk/DCP-LETT-1425.

9. Pauline Boss and Patty Wetterling, "Disappearance, Not Death: The Ambiguous Loss of a Missing Sibling," in Brenda J. Marshall and Howard R. Winokeur, eds., *Sibling Loss across the Lifespan* (London: Routledge, 2017), 175.

10. "Incidence of Twins by Twin Type," Twinsmagazine.com, n.d., at https://twinsmagazine.com/incidence-of-twins-by-twin-type.

11. Anne Casselman, "Identical Twins' Genes Are Not Identical," *Scientific American*, April 3, 2008, https://www.scientificamerican.com/article/identical-twins-genes-are-not-identical.

12. See Archer, *Nature of Grief*, 218–219, and his comments on twin studies.

13. Helen Macdonald, *H Is for Hawk* (London: Jonathan Cape, 2014), 49.

ELEVEN

Life Partners

1. "The Holmes-Rahe Life Stress Inventory," adapted from Thomas Holmes and Richard Rahe, "Holmes-Rahe Social Readjustment Rating Scale," *Journal of Psychosomatic Research* 2 (1967), https://socialwork.buffalo.edu/content/dam/socialwork/home/self-care-kit/holmes-rahe-life-stress-inventory.pdf.

2. Julian Barnes, *Levels of Life* (New York: Knopf, 2013), 86–87.

3. See the Wikipedia entry "Red sky at morning," at https://en.m.wikipedia.org/wiki/Red_sky_at_morning.

4. Kyle J. Bourassa, Lindsey M. Knowles, David A. Sbarra, and Mary-Frances O'Connor, "Absent but Not Gone: Interdependence in Couples' Quality of Life Persists after a Partner's Death," *Psychological Science* 27, no. 2 (2016): 270–281.

5. Megan Devine, *It's Ok That You're Not OK: Meeting Grief and Loss in a Culture That Doesn't Understand* (Boulder, CO: Sounds True, 2017), xv; Megan Devine, "Her Partner Drowned at 39. She Learned That for the Young and Unmarried, Death Has No Playbook," *Washington Post*, August 7, 2018, https://www.washingtonpost.com/news/inspired-life/wp/2018

/08/07/her-partner-drowned-at-39-she-learned-that-for-the-young-and -unmarried-death-has-no-playbook.

6. John Archer, *The Nature of Grief: The Evolution and Psychology of Reactions to Loss* (London: Routledge, 2003), 239.

7. Besell Van der Kolk, *The Body Keeps the Score* (New York: Viking, 2014).

8. Joan Didion, *The Year of Magical Thinking* (New York: Knopf, 2005), 10, 3.

9. Ibid., 10–13.

10. Ibid., 6–7, 33.

11. Antoine Leiris, *You Will Not Have My Hate* (New York: Penguin, 2016), 116.

12. Ibid.

13. "The Widow of Windsor: A Queen in Mourning," Royal Central online, October 24, 2015, at http://royalcentral.co.uk/blogs/the-widow-of-windsor -a-queen-in mourning-55526.

14. Helen Macdonald, *H Is for Hawk* (London: Jonathan Cape, 2014), 199.

15. T. H. White, *The Goshawk* (New York: New York Review of Books, 2007), 15.

16. John Keats, "Ode on Melancholy," https://www.poetryfoundation.org /poems/44478/ode-on-melancholy.

17. J. William Worden, *Grief Counseling and Grief Therapy* (New York: Springer, 2018), 148 (Worden cites the term used by Zisook and DeVaul, which refers to those physical symptoms experienced by the survivor that are similar to those of the deceased); Darian Leader, *The New Black: Mourning, Melancholia and Depression* (London: Penguin, 2009), 50.

18. Leader, *New Black*, 51.

19. Leiris, *You Will Not Have My Hate*.

20. C. S. Lewis, *A Grief Observed* (New York: Harper One, 1994), 12.

21. Ibid., 26.

22. Ibid., 51.

23. Barnes, *Levels of Life*, 74.

24. Lewis, *Grief Observed*, 59. Emphasis in the original.

25. Ibid., 60.

26. Ibid.

27. Barnes, *Levels of Life*, 116–117.

28. Jane Brody, "Recovery Varies after a Spouse Dies, *New York Times*, September 27, 2016, D5.

29. Worden, *Grief Counseling and Grief Therapy*, 77–78.

Epilogue

1. John Matson, "Fact or Fiction?: Lead Can Be Turned into Gold," *Scientific American*, January 31, 2014, https://www.scientificamerican.com/article /fact-or-fiction-lead-can-be-turned-into-gold. Matson describes a 1980 experiment in which a particle accelerator smashed high-speed atomic particles into bismuth, a metal only slightly different from lead (bismuth

has one stable isotope rather than the four found in lead), producing a minuscule amount of gold. The senior author on the 1981 research paper was Glenn Seaborg, a 1951 Nobel winner who shared the prize in chemistry for his work in heavy elements. According to Matson, Seaborg told the Associated Press that it would cost more than a quadrillion dollars per ounce to produce gold by this experimental method.

2. See the Wikipedia entry "Portia (*The Merchant of Venice*)" at https://en .wikipedia.org/wiki/Portia_(The_Merchant_of_Venice).

3. Elysa Gardner, "Diana Krall on Handling Grief, and 'Finding Romance in Everything,'" *New York Times*, June 19, 2017, https://www.nytimes.com /2017/06/19/arts/music/diana-krall-turn-up-the-quiet-interview.html.

4. Ibid.

5. See the Wikipedia entry "Rumpelstiltskin" at https://en.wikipedia.org/wiki /Rumpelstiltskin.

6. Henry Marsh, *Admissions: Life as a Brain Surgeon* (New York: Thomas Dunne Books, 2017), 173.

7. Julian Barnes, *Levels of Life* (New York: Alfred A. Knopf, 2013), 128.

8. Ann Patchett, "Finding Joy in My Father's Death," *New York Times*, February 27, 2015, https://opinionator.blogs.nytimes.com/2015/02/27/ finding-joy-in-my-fathers-death.

9. Ibid.

10. See "Ariel's Song," from William Shakespeare, *The Tempest*, at PotW.org, http://www.potw.org/archive/potw190.html.

11. Helen Macdonald, *H Is for Hawk* (London: Jonathan Cape, 2014), 268. Emphasis in the original.

12. Darian Leader, *The New Black: Mourning, Melancholia and Depression* (London: Penguin, 2009), 202.

13. Ibid., 206. Emphasis in the original.

14. Barnes, *Levels of Life*, 126–127.

15. Roger Rosenblatt, *Kayak Morning: Reflections on Love, Grief and Small Boats* (New York: Ecco, 2012), 146. Rosenblatt's first book is *Making Toast* (New York: HarperCollins, 2010).

16. Max Porter, *Grief Is the Thing with Feathers* (Minneapolis: Graywolf, 2015), 109–110, 7.

17. John M. Marzluff and Tony Angell, *In the Company of Crows and Ravens* (New Haven: Yale University Press, 2005), 111, 289–290.

18. Jennifer Holland, "Wild Messengers," *New York Times*, November 1, 2014, https://opinionator.blogs.nytimes.com/2014/11/01/wild-messengers.

19. Kenneth J. Doka, *Grief Is a Journey: Finding Your Way through Loss* (New York: Atria, 2016), 251. Doka cites Drs. Lawrence Calhoun and Richard Tedeschi, who describe some of the ways that growth can result from grief.

Credits

Index